Schriftenreihe des Energie-Forschungszentrums N

Band 57

Das EFZN ist eine wissenschaftliche
Einrichtung der

TU Clausthal

in Kooperation mit den Universitäten

Technische
Universität
Braunschweig

GEORG-AUGUST-UNIVERSITÄT
GÖTTINGEN

Leibniz
Universität
Hannover

CARL
VON
OSSIETZKY
universität OLDENBURG

Incentives to value the dispatchable fleet's operational flexibility across energy markets

Doctoral Thesis
(Dissertation)

to be awarded the degree

Doctor of Engineering (Dr.-Ing.)

submitted by

Eglantine Künle
from Fribourg

approved by the Faculty of Energy and Management,
Clausthal University of Technology,

Date of Oral examination
14.09.2018

Dean
Prof. Dr.-Ing. Norbert Meyer

Supervising tutor
Prof. Dr.-Ing. Martin Faulstich

Reviewer
Prof. Dr.-Ing. Richard Hanke-Rauschenbach

Bibliografische Information der Deutschen Nationalbibliothek
Die Deutsche Nationalbibliothek verzeichnet diese Publikation in
der Deutschen Nationalbibliografie; detaillierte bibliographische
Daten sind im Internet über http://dnb.d-nb.de abrufbar.
1. Aufl. - Göttingen: Cuvillier, 2018
   Zugl.: (TU) Clausthal, Univ., Diss., 2018

D104

© CUVILLIER VERLAG, Göttingen 2018
   Nonnenstieg 8, 37075 Göttingen
   Telefon: 0551-54724-0
   Telefax: 0551-54724-21
   www.cuvillier.de

   ISBN 978-3-7369-9890-2
   eISBN 978-3-7369-8890-3

À mon grand-père et à tous ceux qui me sont chers

Table of content

Abstract

To keep our planet livable, hundred and seventy-five nations worldwide, including Europe, have committed to drastically reducing greenhouse gas emissions. The work at hand participates with the energy transition toward decarbonized electricity production by providing a method to analyze the incentives for the thermal fleet to contribute. Given the liberalized economic environment Europe and other countries have put in place, the work at hand analyzes the signals set by the energy markets to investors and power plant operators regarding flexible power plant operation and the future requirement of power plant flexibilisation, with improvement suggestions.

This work contributes to the field of energy system analysis and operations research. First, a methodological framework for quantitative, long-term, high-resolution techno-economic assessments of an individual power plant's operational flexibility is put in place. Instead of relying on single metrics, this method uses the power plant's unit commitment and economic dispatch for flexibility valuation. Depending on the desired insight, different metrics might be derived from the plant dispatch. The single unit, self-scheduling problem of a merchant power plant solved from a price-taker perspective developed in this work includes the intraday and frequency control markets, as well as a parametric power plant model suited for operational flexibility assessments. The model accounts, among others, for the degraded power plant efficiency at part-load. In addition, significant aspects, like the variation of start-up costs as a function of the market prices are included. Improved opportunity cost calculation methods including part-load efficiency degradation for the optimization of capacity reservation in the frequency control market are developed. Quantitative analyses demonstrate that comprehensive flexibility studies should include intraday and frequency control markets. The developed concepts might also be implemented in models solving other types of unit commitment problems than the single unit self-scheduling problem. The proposed formulations are suited for the state-of-the-art mixed integer programming approach. The work at hand, however, introduces an alternative approach. Acknowledging that the power plant operation is a controlled process with a finite number of states and transitions, the concept of events is used to overcome uniform time discretization. This event-based approach requires less computation time than time-discrete approaches when simulating long time horizons.

The work's practical relevance is illustrated in the retroactive observation of how the merchant power plants react to the markets in their operation strategies. The case studies show that the day-ahead markets incent a conventional power plant's flexible operation at increased shares of renewables, but currently do not compensate for the profit difference resulting from this system-friendly dispatch. The analyzed frequency control and intraday markets help in improving the day-ahead profit and monetary value of power plant flexibilization. The design of the frequency control capacity reservation market is shown to benefit more power plants (renewables as well as conventional ones) when the capacity reservation duration is short. The value of operational flexibility improvements is quantified using various metrics, for different technologies and market environments. These case studies indicate that if investors further expect a three to five-year return on investment, the existing conventional fleet is unlikely to experience a significant flexibilization.

Kurzfassung

Um den Planeten lebenswert zu halten, haben sich weltweit hundertfünfundsiebzig Nationen, und insbesondere die Europäische Union, dazu verpflichtet, die Treibhausgasemissionen drastisch zu reduzieren. Die vorliegende Arbeit untersucht den Beitrag thermischer Kraftwerke zum Aufbau einer dekarbonisierten Stromerzeugung. Vor dem Hintergrund der Liberalisierung der Energiewirtschaft in Europa und anderen Ländern werden die bestehenden Anreize der Energiemärkte für Investoren und Kraftwerksbetreiber zum flexiblen Betrieb der Kraftwerke, sowie ergänzende Maßnahmen zur Kraftwerksflexibilisierung analysiert und Optimierungen vorgeschlagen.

Es wird ein methodischer Rahmen für eine quantitative, langfristige und hochaufgelöste technoökonomische Bewertung der betrieblichen Flexibilität eines einzelnen Kraftwerks vorgeschlagen. Anstatt sich auf eine Metrik zu konzentrieren, verwendet die hier vorgeschlagene Methode eine Kraftwerkseinsatzplanung zur Flexibilitätsbewertung. Abhängig von der gewünschten Erkenntnis können unterschiedliche Metriken aus dem Kraftwerkseinsatz abgeleitet werden. Das in dieser Arbeit entwickelte, aus einer Preisnehmerperspektive herausgelöste Problem des Einsatzes eines Kraftwerks, umfasst die Intraday- und Regelleistungsmärkte sowie ein parametrisches Kraftwerksmodell, das für betriebliche Flexibilitätsanalysen geeignet ist. Das Modell berücksichtigt unter anderem den reduzierten Kraftwerkswirkungsgrad bei Teillast. Darüber hinaus sind wesentliche Aspekte wie die Veränderung der Anfahrkosten in Abhängigkeit von den jeweiligen Marktpreisen enthalten. Verfeinerte Verfahren zur Berechnung der Opportunitätskosten, einschließlich einer Verschlechterung des Teillast-Wirkungsgrads für die Optimierung der Kapazitätsvorhaltung auf dem Regelleistungsmarkt, werden vorgestellt. Quantitative Analysen zeigen, dass Berechnungen ohne Intraday- und Regelleistungsmärkte gegebenenfalls den Wert der betrieblichen Flexibilität unterschätzen würden. Die in der vorliegenden Arbeit entwickelten Konzepte sind für alle Betrachtungsweisen der Kraftwerkseinsatzplanung geeignet. Diese Konzepte sind in der üblichen zeitdiskreten Formulierung der gemischt-ganzzahlige Kraftwerkseinsatzoptimierung implementierbar. Ein darüberhinausgehender Ansatz wurde vorgeschlagen. Da der Kraftwerksbetrieb ein kontrollierter Prozess mit einer endlichen Anzahl von Zuständen und Übergängen ist, wird ein ereignisbasierter Ansatz verwendet, um die Notwendigkeit einer einheitlichen Zeitdiskretisierung zu überwinden. Der hier gewählte Ansatz erfordert gegenüber dem zeitdiskreten Ansatz weniger Rechenzeit bei der Simulation langer Zeiträume.

Die praktische Relevanz der vorliegenden Arbeit liegt in der auf historischen Datensätzen basierten Simulation, wie Kraftwerke in ihren Betriebsstrategien auf die Märkte reagieren. Fallstudien zeigen, dass die Day-Ahead-Märkte den flexiblen Betrieb eines konventionellen Kraftwerks bei erhöhten Anteilen erneuerbarer Energien fördern, aber derzeit die Gewinndifferenz, die sich aus diesem systemfreundlichen Betrieb ergibt, nicht kompensieren. Die analysierten Regelleistung- und Intraday-Märkte tragen dazu bei, den Day-Ahead-Gewinn und den Geldwert der Flexibilisierung von Kraftwerken zu verbessern. In der vorliegenden Arbeit wird gezeigt, dass der Regelleistungsmarkt mehr Kraftwerke (sowohl erneuerbare als auch konventionelle Kraftwerke) nutzt, wenn die Kapazitätsreservierungsdauer kurz ist. Der Wert der Verbesserungen der betrieblichen Flexibilität wird anhand verschiedener Metriken für verschiedene Technologien und Marktumgebungen quantifiziert. Diese Fallstudien deuten darauf hin, dass der bestehende Park konventioneller Kraftwerke wahrscheinlich keine signifikante Flexibilisierung erfahren wird, wenn die maximale Amortisationszeit der Investoren weiterhin drei bis fünf Jahre betragen darf.

List of illustrations

List of tables

Part A

Theoretical work foundation

1 Motivation

To keep our planet livable, hundred and seventy-five nations worldwide, including Europe, have committed to limiting the global temperature increase by drastically reducing anthropogenous greenhouse gas emissions. The current resource we are relying on globally to produce electricity and heat is the most carbon-intensive one: coal. In 2015, coal accounted for 39.3% of the world's electricity production and for 44.9% of the world's CO_2 emissions from fuel combustion[1]. Electricity and heat accounted for 42% of the global CO_2 emissions in 2015 [2]. The energy sector's logical contribution to climate protection targets is thus to reduce, and ultimately eliminate, the use of greenhouse gas emitting fuels for electricity production in favor of clean ones. This transition is a very challenging (and crucial) one. The author wishes to concur to the success of this transition by providing a method to analyze the incentives for the thermal fleet to contribute.

The most apparent challenge raised by the energy transition when relying on variable renewable energy sources is the intermittency they introduce in the system, and thus the increasing need for operational flexibility. Given the liberalized economic environment Europe and other countries have put in place, the work at hand analyzes the signals set by the energy markets to investors and power plant operators regarding flexible power plant operation and the future need for plant retrofitting. The author aims to put in the hands of everyone (either the ones who believe that there should be absolutely no incentive for these plants to remain or the ones who believe that we will need them for a long time) a method to quantitatively assess the incentives for flexible operation.

This thesis is divided into three parts: the theoretical work foundation, the development of the methods and concepts and finally their application. Each part is divided into sections, at the beginning of which the subsections are detailed. Part A section 1 introduces the research background and problem. Part A section 2 compares the review of existing problem-solving approaches with the specification of a suited approach. This comparison defines the research gaps, and thus the research questions that will be answered. Part A section 3 introduces key concepts from the literature.

1.1 Energy system decarbonization

Two years after its creation in 1988, the Intergovernmental Panel on Climate Change (IPCC) provided the first report evidencing that human activities cause the increase in greenhouse gas concentration in the atmosphere, and thus climate change. Two years later, the United Nations Framework Convention on Climate Change (UNFCCC) was adopted and came into force in 1994. Since then, the United Nations Framework Convention on Climate Change (UNFCCC) has been the primary international forum for resolutions against climate change. In 1995, it was established that the commitments taken in Rio were not ambitious enough, leading to the Kyoto protocol adoption two years later. In 2005, the Kyoto protocol came into force and one of its pillars, the European trading system, now regulating half of the worldwide CO_2 emissions, was put in place. The second pillar, the clean development mechanism, was put in place in 2006. This mechanism offsets emissions in industrialized countries with sustainable projects in developing countries. The second commitment period to the Kyoto protocol was agreed upon in Doha in 2012. Climate change and its consequences have been evidenced again and detailed in the IPCC's Fifth Assessment Report, which addressed the need for action against climate change. In 2015 a universal agreement for the period beyond 2020, the Paris Agreement, was signed by 195 nations. Each nation submitted its contribution to the goal of maintaining the temperature increase well below 2°C compared to pre-industrial times. These commitments can be visualized using the Paris Reality Check developed in the PRIMAP group at the Potsdam Institute for Climate Impact Research [3]. The European Union, and thus its collaborating member states, has committed to a binding target of at least 40 % domestic reduction in greenhouse gas emissions by 2030 compared to 1990 [3].

As greenhouse gas emissions mainly stem from the combustion of fossil fuels, the energy sectors for electricity and heat production, transportation, and industry are crucial to reaching the binding targets of the Paris Agreement [4]. The European Union addresses this problem by increasingly relying on emission-free resources for electricity production and improving energy efficiency. The focus of the present work is on the electricity and heat production sector, which has seen a significant increase in renewables share, reaching 17 % of the European energy consumption and 29.6% of the European electricity production in 2016 [5]. Among the set of possible technologies, a majority of the European countries which try to decarbonize their energy system place their hopes in volatile resources like wind and sun. Alternative or complementary emission-free energy production is based on hydropower, carbon capture and storage or utilization at conventional plants, biomass firing, and nuclear power. The national and local energy mix depends on the natural resources and past as well as present political decisions. A further crucial aspect is the social acceptance of the energy mix, as shown by the German nuclear exit. However, this aspect will not be addressed in this work. Regarding feasibility, some studies have shown that fulfilling the Paris Agreement based on renewable energy sources is manageable, see, e.g., [6], [7], [8]. Some of these studies are assessed in [9]. While some reports indicate that a coal exit is required by 2050 [10, 11] or before [12] to meet the Paris agreements, Italy, the United Kingdom, the Netherlands, Ireland, and others have decided to phase out coal. Many governmental organizations and also businesses divest from coal, see, e.g., [13]. The increase in renewables is thus concomitant to the reduction of fossil-based electricity production, but at different paces [14]. The authors of [15] report the new policy projections of the International Energy Agency's world energy outlook in 2040: the power generation would rely 27 % on oil, 24 % on gas and 23 % on coal. The energy systems during the transition toward a fully de-carbonized system are thus likely to rely on a mix of fossil and renewable-based electricity and heat production.

1.2 The operational flexibility challenge

It is thus likely, that a growing share of variable renewables will be integrated into a system designed for dispatchable power generation technologies. The primary physical challenge will be the integration of variability within the grid to further ensure adequacy between demand and supply at any time. During the energy transition, dispatchable electricity generation and sinks will be relied on to integrate renewables. The primary requirement formulated in the literature is therefore flexibility. An overview of flexibility definitions is given before reviewing the possible generation technologies and sinks providing flexibility.

1.2.1 Overview of flexibility definitions

A review of flexibility definitions that can be found in the field of power system analysis is provided in Table 1. It shows that flexibility qualifies the response of the system or its units to a variable signal. This qualification uses the parameters time, speed, costs, system state, uncertainty, reliability and even location in [16]. From a transmission network perspective, flexibility is the capacity to adapt to a new environment at minimum cost (and time), see, e.g., [17], [18], [19], [20]. Flexibility is often measured as the ramping capability of some units or the system as a whole ([21], [22]). The authors of [23] define the concepts of flexibility and adaptability to characterize change, independently of the nature of the system to which change applies. The authors use flexibility to characterize a change initiated by an agent external to the system, whereas adaptability is used to characterize a change initiated by an agent internal to the system [23]. The comparison of two systems' flexibility is thus based on the number of ways a system can change from one state to another within an acceptable cost threshold. To determine whether a system is flexible or not is to determine whether a change from one state to the other can be done within the desired cost threshold [23].

1.2.2 The operational flexibility providers

The following references have been used to gather a comprehensive review of possible flexibility providers and means: [24], [25], [26], [27], [28], [29]. Flexibility providers are not only available in the electricity sector but also in the heat and mobility sectors. Denmark, as an example, shows that coupling the electricity and heat sectors via, e.g., combined heat and power plants offers a way to integrate significant amounts of intermittent renewables. This integration requires power plants with flexible electricity and heat production (means extraction plants should be preferred to back-pressure plants, see, e.g., [26]). Heat and power can further be decoupled with heat storage technologies on site. In most of the current energy systems, the existing fleet relies on thermal power plants. These plants are controllable and reliable but might lack operational flexibility. The dynamically responding conventional power plant fleet can be seen as the main flexibility provider during the energy *transition*, but not in the targeted system, as these plants are significant greenhouse gas emitters and thus do not fit the environmental friendly requirement. Technical solutions for the capture of CO_2 are available, but the question of storage and utilization remains. Even nuclear power plants offer flexible ramp-rates as demonstrated in France. Renewable generation can be curtailed or modulated downwards, offering flexibility in one direction.

Regarding demand, the International Energy Agency distinguishes between managed and responsive resources in [24]. The managed response is contracted in advance whereas the responsive resources react dynamically to the system needs. Between demand and supply, storage offers a way to displace production and consumption. Its flexibility depends on the used technology and is measured in terms of capacity, charge and discharge speed. Power storage technologies can be based on mechanical, chemical and electrochemical processes. Power can be stored in the form of heat, gas or mobility, before being transformed again into power. Stationary storage capacities and time-variant storage capacities are differentiated in [29]. Hydro storage, Compressed Air Energy Storage (CAES), flywheel systems, redox-flow cells, advanced capacitors, superconducting magnetic energy storage belong to the first category. The electric vehicle fleet belongs to the second category.

The grid is crucial to enabling flexibility, as sufficient and available flexibility resources are of no use without grid support. Similarly, proper interconnections between grid zones allow for sharing flexibility. Larger balancing areas offer more geographic inhomogeneity, helping to smooth the effects of increased share of variable renewables. The power imports and exports facilitated by the grid also require appropriate market design and improved operations. The next section addresses this topic in the European context.

Table 1 Review of flexibility definitions in the field of energy systems analysis

Flexibility definition	Source
"A flexible plan is one that enables the utility to quickly and inexpensively change the system's configuration or operation in response to varying market and regulatory conditions."	[30]
"System flexibility is defined as the ability of supply-side and demand-side resources to respond to system changes and uncertainties. Flexibility also includes the ability to store energy for delivery in the future and the operational flexibility to schedule/dispatch resources in the most efficient manner."	[31] [32]
"[...] the potential for capacity to be deployed within a certain timeframe."	[21]
"Flexibility expresses the extent to which a power system can modify electricity production or consumption in response to variability, expected or otherwise. In other words, it expresses the capability of a power system to maintain reliable supply in the face of rapid and large imbalances, whatever the cause. It is measured regarding megawatts (MW) available for ramping up and down, over time."	[24]
"[...] technical capability of individual power system units to modulate power and energy in-feed into the grid, respectively power out-feed out of the grid." Same measure as in [33]	[29, 34]
"[...] the ability of a system to deploy its resources to respond to changes in net load." Measured as the "expected number of observations when a power system cannot cope with the changes in net load, predicted or unpredicted"	[35]
"[...] system's capability to respond to a set of deviations that are identified by risk management criteria through deploying available control actions within predefined time frames and cost thresholds."	[36] based on [37] and [23]
"[...] the ability of a power system to cope with variability and uncertainty in both generation and demand, while maintaining a satisfactory level of reliability at a reasonable cost, over different time horizon."	[22]
"Technical flexibility of a component according to the allowable disturbance magnitude at the component of choice."	[38]
"Operational flexibility describes a power system's ability to respond with controllable real power resources to rapid changes in power balance error."	[39]
"The flexibility of a given system is a unique, innate, state and time-dependent quality. In conversation, it is therefore sometimes said that flexibility is the ability to deviate from the plan."	[40]
"Flexibility is the ability of a power system to respond to changes in power demand and generation," measured as "the magnitude and frequency of net load ramps of given duration that have to be balanced by the complementary system."	[41, 42, 42]
"Operational flexibility is the ability of a power system to contain a disturbance sufficiently fast in order to keep the system secure," "Locational flexibility is the operational flexibility that can be accessed at a given bus in the grid. It describes the disturbance at a given node of the system that could be contained by suitable and available remedial actions."	[16]
"Flexibility at a given state is the ability of a system to respond to a range of uncertain future states by taking an alternative course of action within acceptable cost threshold and time window,"	[43]

1.3 Liberalized market environment

The primary components of the energy system are generation, transmission, distribution, and retail supply, and are historically organized as vertical monopolies [44]. Parallel to the efforts for a sustainable future, some energy systems worldwide have gone through liberalization to increase competition for welfare maximization. Wholesale and retail markets have been introduced to shift risk and costs from consumers to suppliers [44]. The wholesale market should incent generators to invest in the right technologies and system operation in order to provide adequate service [44]. A review of the liberalization process and the different degrees of achievement is provided in [44]. In fact, there are still many disparities in the implementation of the electricity markets. A good example is the inclusion or exclusion of capacity markets, or price formation including start-up costs. Moreover, as renewables are subsidized for their preferred feed-in, these only partially compete with other generation technologies. In Europe, however, there is a move from feed-in-rates toward project tendering [45]. This section provides an overview of the functioning of the European physical electricity markets, schematically illustrated in Figure 1. Since the North-American electricity market uses different approaches, it will also be explicated for comparison. The terms scheduling and dispatch are used in the following. Scheduling refers to the decision to operate units or not. Once the schedules have been defined and submitted, an entity nominates the schedules. The nominated entities then proceed to the dispatch, which defines the generating level of the nominated units.

The liberalized power system comprises five primary actors: generation companies, system operators, market operators, retailers, and consumers. In a centralized market design, the market clearing and transmission system operation are performed at the same time by the same entity (centrally), whereas in decentralized markets, both duties are separate [46]. The market operator organizes energy and cross-border capacity trading, whereas the system operator is responsible for the secure operation of the network. The market participants schedule and dispatch themselves and declare their dispatch to the system operator. In case of system imbalances, the latter has the responsibility to send dispatch instructions. The main difference between both systems is the system operator's responsibility. In the central approach, the system operator is the only responsible party, whereas, in the decentralized approach, other parties, such as the market operator or balance responsible parties have been defined. The decentralized design can be supplemented by an organized market for ancillary services, where the system operator contracts services from generators. In Europe, most of the energy systems are decentralized, and central dispatch is in place in few countries like Greece, Ireland, Italy and Poland, see [47]. A further hybrid approach uses self-scheduling and central dispatch [48].

In the case of Europe, the system operator is also the transmission system operator (TSO). In the United States, the independent system operator (ISO) is in charge of system operation. In this case, the transmission system owner is not the system operator, as is the case in Europe. In Europe, most of the national electricity markets have converged to a system with forward markets, short-term markets at day-ahead and intraday time frames, and organized markets for frequency control provision. Forward markets are intended for long-term hedging, whereas day-ahead and intraday markets are in place to form the short-term electricity wholesale price. Two price formation mechanisms are possible [49]: pay-as-bid (which means that all cleared parties are paid their bid price), or single pricing (which means that all actors get the same price). With enough market liquidity and supply and demand elasticity, the price formed by the intersection of demand and supply should be at its competitive level [50]. The day-ahead market time discretization is often hourly, and organizes the trading for the next day, either bilaterally or via power exchanges. The intraday markets organize the trades within the day to correct forecast errors. The settlement might be organized either as a fixed gate auction or as a continuous-time auction. The first design sets a time at which bid submitting stops, and the market is then cleared. The continuous design is cleared whenever a sell order matches a bid order. The design of the markets is often distinguished between the formats of the offers that can be placed by the actors, referred to as simple, block or complex orders [48]. The two first formats are in place in most of the European countries, while complex offers are in place in the United States and the Iberian market [48]. The latter allows, e.g., start-up costs to be accounted for in the offer. Block orders allow conditional submissions, with exclusion or linking of adjacent hours. In Europe, the only ancillary service organized within a market is for grid frequency control. This market is divided into the reservation of capacity and the consequent eventual delivery of the balancing energy via the generator's load increase or decrease. In Europe, the reserves and energy are cleared sequentially, while these are co-optimized in the United States.

After the electricity markets' liberalization, in addition to energy system de-carbonization, Europe is working at becoming an integrated electricity market to maximize the region's welfare, see the European Union's third energy package (Directive 2009/72/EC for the internal electricity market). The European Network of Transmission System Operators for electricity (entsoe) is helping to organize the transition toward the internal energy market, with the primary role of cross-border transmission allocation optimization, see the European Union's third energy package (Regulation (EC) 714/2009).

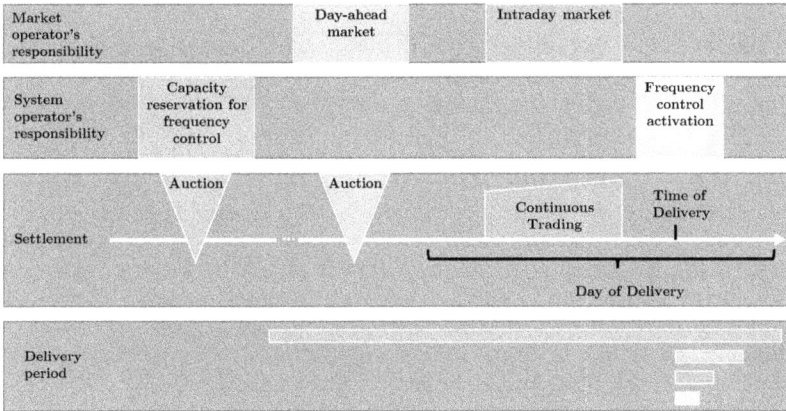

Figure 1 Functioning of the physical markets in Europe. The design and time frames might vary from one country to another. This representation captures the most used implementation schematically.

1.4 Research question

The previous sections have shown that the aim of the energy transition is a sustainable and competitive system, ensuring the security of supply. Not only the variations of power but also the speed of these variations need to be known to ensure the security of supply and reduce renewables' integration costs. The volatility of renewables-based supply makes accurate forecasting more difficult. Forecast errors lead to even more intermittent operation of the dispatchable fleet, which had not been designed for this mode of operation. The role of conventional power plants in the energy transition, in light of the facts exposed in the previous sections, is to be a flexible provider to the demand that cannot be met by renewables (referred to as the residual load). Other flexibility options to dispatchable backup capacity such as electricity storage and demand side management might relieve the conventional power plants from this task as these reach the end of their lifetime. In the current European context, there will be few or no investments in new thermal power plants, but retrofits for compliance to regulatory constraints or to sell (and make money with) flexibility might come into question. In the field of power plant technology, retrofits are power plant component or subsystem improvements [51], and might be hardware as well as software. In this regard, it is worth knowing what a good flexibility improvement from a merchant power plant operator and investor's point of view is.

The role of dispatchable power plants as flexible residual load providers in a liberalized context raises the question whether these plants are technically able to provide the required flexibility but, even more, if the market environment incents the required flexible dispatch. Loop flows between Germany and its neighbors have led to the installation of phase shifters and finally to the splitting of the German and Austrian common price zone. This example shows that not only technologies but also the design of the markets and the way operation is planned, especially in a liberalized environment, are enablers for the renewables integration. The challenges in this regard might be distinguished between planning and operation. According to, e.g. [52] and [49], a concern raised by the current market designs is whether these set the right investment incentives. Even if there is currently no capacity shortage in Europe [53], the authors of [52] suggest that a market sending no investment signal might lead to shortages in the future. To address this planning issue, capacity markets have been widely discussed and implemented in various forms in Europe. See, e.g., [54], [49] for discussion and review of capacity mechanisms. Additionally to this planning perspective, many studies assess the need for flexibility at the operational time scales (see review in section 2.1), but how can it be ensured that the incentives are set up to encourage flexible operation and eventually investment in the required flexibility in a liberalized context? Are the markets well designed for flexibility incentives?

A quantitative assessment of the value of operational flexibility at thermal power plants in their market environment is required to answer these questions. This work aims at developing the methods to perform this quantitative assessment and to provide the tools for answering the above questions at given power plants in given market contexts. From a methodological point of view, it clarifies how to quantify the value of operational flexibility. The next section specifies such a methodology. A review of available methodologies is used to define the research gaps and contributions of this work.

2 State-of-the-art methodologies and research gap

2.1 Flexibility in the literature - a review of assessment methodologies

When assessing flexibility, two main tasks can be distinguished: quantify the need for flexibility and assess the available flexibility to state about its adequacy. From a system perspective, a global flexibility assessment would compare a quantified flexibility need with the available flexibility. The International Energy Agency defined such a holistic approach in its FAST tool [24], in which the estimation of the system's amount of flexibility allows to state about the potential variable generation penetration. However, the approach is based on a stylized representation of the resources without considering their availability, raising the need for more detailed assessment for firm decision drawing. The literature about operational flexibility is mainly divided into system planning and system operation [55], and thus to the respective time horizons. Power system studies for planning are further divided into transmission expansion planning and generation expansion planning [37]. This subsection provides a review of this literature.

2.1.1 System flexibility requirements

In system planning, metrics have been developed to assess the system reliability, as, e.g., the loss of load expectation (LOLE). See [56] for more reliability metrics. Hence, as stressed in [57], similar metrics could be developed to assess the system flexibility. In [35] it is aimed at identifying how well a system can cope with the overall variability of the net load. The authors do so by developing the insufficient ramping resource expectation metric (IRRE), which is derived from traditional generation adequacy metrics such as LOLE. This metric assesses the expected number of net load ramps for which insufficient ramping resources would be available. This approach is aimed at long-term system planning and does not break down to the assessment of a single plant or technology. The IRRE has been completed by the periods of flexibility deficit (PFD), the expected unserved ramping (EUR) and well-being metrics in [58] (see Table 2) and implemented in the Electric Power Research Institute's InFLEXion tool. Due to the tremendous needs for system modeling, this tool's implementation does not account for the system dispatch but only on likely commitment status. Similarly, in [59] the authors define the lack of ramp probability (LORP), a variation of the loss of load probability (LOLP), which calculation also requires the whole generation units' dispatch status to be known. This metric is intended to measure the risk of ramping shortage attributed to a dispatch decision in the operational timescale. The work [60] provides an assessment of several performance indices for the evaluation of operating reserve requirements in long-term planning, relying on a scheduling algorithm based on a Markov-chain approach.

In [33], the authors quantify the increased need for load following and regulation reserve in real time and hour-ahead markets. The authors calculate the flexibility requirements from an operational perspective via a dispatching tool. For the Statistical Analysis of the Regulation and Load Following Requirements, three decision variables have been introduced. These are the regulation and load following capacity, ramping speed and ramp duration. Based on the work of [33], [61] formulates an optimization problem to obtain the day-ahead energy and reserve procurement meeting the system flexibility requirement at a given probability level. This measure is based on magnitude, ramp-rate and duration of net load deviations from their scheduled values. [62] and [61] assess the system reserve requirement via the net load variability resulting from the renewables share. The first reference determines the net load based on a fundamental model, whereas the second makes use of historical data. Similarly, the authors of [42] provide statistical analysis of the load variability regarding ramp-rate frequency and duration for some European countries at operational timescale. [63] uses a macroeconomic approach also based on a fundamental model, optimizing the capacity and dispatch of flexibility options minimizing the system costs for a given load and renewables share. The flexibility need is defined as the residual load that needs to be provided by dispatchable sources. The approach in [64] is the same, but with a more comprehensive fundamental model including the interconnection capacities between countries. The flexibility need is in this instance defined as the difference between the residual load and its average over predefined timeframes.

2.1.2 System and technology flexibility adequacy

Assessing the system flexibility requirements is often a preliminary step to the determination of the system flexibility resources adequacy. The latter is of interest to the research problematic. The following section provides a review of this topic in the literature. In [65], a visualization of the existing fleet is proposed to characterize the country's specific available flexibility. This very high-level visualization is limited to the penetration of interconnection capacities, hydro capacity, pump storage capacity and combined heat and power plants capacity compared to the required peak capacity. The reference [25] provides a review and clustering of methods for assessing the potential of flexibility options, with focus on storage capacity requirements. The approaches are classified into approaches assessing the flexibility requirement from a technical, economic or market perspective. The last category is the less represented one and therefore requires more extensive research.

2.1.2.1 Flexibility metrics

In [29] and [34], the authors develop a metric for the technical capability of individual power system units to deviate from their schedule. The metric is composed of three parameters, which represent the maximum flexibility capability of the power plant, which limits its actual power trajectory. These three parameters are referred to as the flexibility trinity and are the basis of the flexibility metrics presented in this section. They are the power ramp-rate ρ, power π, and energy ε. [16] uses a similar metric, which it also applies to system nodes and goes further by defining the concept of locational flexibility, and proposing a two-stage robust optimization for the procurement of control reserves. [38] also develops a quantitative method for single component analysis but with *a priori* specified reliability criteria, linking a generator model and external power disturbances. The generator model is not the focus of the presented publications, so these lack the technical detail regarding operational flexibility specific topics like efficiency losses at part-load. In [39], the authors address short-term operational flexibility with a metric quantifying the real power resources, which can be redeployed within a given timeframe without violating any constraints at a given certainty level. This metric is intended for very short-term scheduling decisions. The reference [43] proposes a unifying metric, comparing the most significant variation range of uncertainty with the target range reflecting the wished risk level. By addressing the aspects of time, action, uncertainty, and costs, the proposed metric can represent the available upward and downward ramping available to the system as well as a score indicating how much of the target uncertainty can be met [43]. [37] proposes to measure flexibility as the proportion of system outcomes (demand and production scenarios) for which an operating strategy guarantees a certain level of risk.

[66] measures the flexibility of a transmission network plan as the variance of the cost-incurring for a set of deviations from the planned situation. [17] defines flexible transmission plans as the ones requiring less cost but also time to adapt to a new unforeseen environment. The flexibility is measured as the adaptation cost of the optimal expansion plan of a scenario to another scenario (to which it is not optimal). In [67] the methodology, rather than choosing the most flexible plan among a set of candidate solutions, is designing a flexible system that is less sensitive to the choice of scenarios by minimizing the future adaptation cost to the conditions of other identified scenarios. [20] assesses transmission expansion plans by measuring the expected unserved demand. [18] measures the transmission network adequacy from planning as well as from an operational perspective via two flexibility indexes. The Technical Uncertainty Scenarios Flexibility Index (TUSFI) and Technical Economical Uncertainty Scenarios Flexibility Index (TE-USFI) are an improvement of [19]. It introduces the uncertain scenarios flexibility index (USFI), which is the measure of the flexibility of a transmission network among several alternative configurations from a planning perspective. The index calculates the average of the load flow distribution factors of the network branches, weighted with the margins of the branches at a given time. The margins are derived from generation scenarios. The authors of [68] optimize the coordination of flexibility between Transmission System Operators, with the exportable flexibility defined as the reserve that can be exported via tie-lines. The dispatch is defined exogenously, and the coordination of the available flexibility between the areas in real-time is optimized. In [69], the authors assess the influence of the transmission network on the flexibility available for balancing. Two aspects are addressed: the influence of the network

constraints and congestion on the available flexibility in real time. Metrics already introduced in the previous section are used to assess the system's flexibility.

2.1.2.2 Levelized cost approaches

The levelized cost of electricity (LCOE), a further metric, is a state-of-the-art approach supporting decision-making when considering investments in new power plant projects. [70] provides a suited introduction to LCOE and its caveats. [70] defines the LCOE as "for a given generation plant, the constant (in real terms) price for power that would equate the net present value of revenue from the plant's output with the net present value of the cost of production." Mathematically, the LCOE supposes the discounting of cash flows and capital costs. When it comes to LCOE calculations, different assumptions or parameters are included in the calculations, as shown in [71], thus making the comparison from one study to the next pretty difficult. The main caveat according to [70] is that, despite being the basis for future investments, the LCOE calculations mainly base on past or recent projects.

LCOE is, however, sometimes referred to when addressing flexibility. The International Energy Agency uses the levelized cost of flexibility (LCOF), a derivate of the LCOE to assess different flexibility technologies [72]. For dispatchable generation technologies, this metric includes capacity factors and cycling regimes. [73] introduces a capacity value in its LCOE calculation to distinguish between dispatchable and non-dispatchable units. To overcome further LCOE caveats, [73] even introduces a further metric, the levelized avoided costs of energy (LACE) to reflect the value of a project via the costs of energy production without the project realization. Other derivatives of the LCOE have been introduced, as the "system LCOE" in [74], since the LCOE does not account for the heterogeneity of power production technologies.

As pointed out in [23], the valuation of flexibility by means of a single metric requires assumptions on how to represent crucial parameters like time and costs, thus introducing a loss of generality. As such parameters are often related, such a metric (flexibility metrics as well as LCOE) would necessarily conceal their interactions [23]. The authors of [27] also note that a single metric does not satisfy flexibility assessments. With a mind to flexibility, the main caveat of LCOE is that it does not include the volatility and dynamics of the market and plant operations, as noted in [75]. As these metrics (LCOE as well as LCOF) do not capture any dynamic, the International Energy Agency needed to complete the LCOF assessment in [72] with models for power system planning and operation. Such approaches are reviewed in the following subsection.

2.1.2.3 System operation

In [76], the authors use a production cost model to assess the value of wind forecasts at different time scales and use metrics to quantify the factors influencing available flexibility. The fleet's static flexibility is defined in terms of the generation mix, must-run capacity, ramp-rates and minimum generation level. The generation fleet's static flexibility is then improved by the dynamic flexibility measured by the ramping capacity at a given dispatch situation. [21] explores the adequacy of flexibility resources to meet the system flexibility requirements by adding a new constraint to the power system's unit commitment problem. This constraint states that the available potential deployable flexibility must be higher than the system flexibility requirement. In this approach, the latter is derived from the wind power variability. The deployable flexibility is the sum of the deployable capacity of all flexible resources of the system, which is based on the expected status and dispatch of the resource. A similar approach is implemented in the InFLEXion tool already mentioned above. In their conclusion, the authors of [21] underline the need for including generating unit cycling in their problem formulation.

[77] proposes a dedicated unit commitment formulation with planning (investment optimization) and operation, thus providing an attempt to combine system planning and operation. The idea of combining unit commitment and expansion planning is pursued in [22] with a dedicated measure of the flexibility of a resource. The flexibility metric is a normalized measure of a power plant's dispatchable range and ramp-rate capability, and the system's flexibility is computed as the weighted sum of the individual flexibility metrics. Unit commitment is used to determine which amount of variable renewable generation can be integrated into a system as a function of its flexibility index. The proposed methodology is, however, missing the intra-hourly variation of flexibility requirements. [78], based on previous work of [33], proposes a unit commitment formulation including flexibility constraints in the form of upward and downward reserve, ramp-rate and ramp duration requirements. This supposes that the three upward and downward flexibility requirements have been assessed previously with a method presented in 2.1.1.

In [79] flexibility is assessed via a multi-criterion ranking of technologies and a power system model, to analyze storage from a system perspective as well as from a market player point of view. The power system model, when used from the system point of view, optimizes the investment in and operation of storage technologies depending on various energy system scenarios, without performing the optimization of the whole energy system. When addressing the market player point of view, the same methodology is applied but with a mind to taxes and duties.

[80] addresses the flexibility of generic technologies from a system planning (investment and idling) perspective. Based on [30], the author defines flexibility as the difference in system performance (measured in terms of costs) with and without uncertainty. In [80], a planning model is used to evaluate the cost of meeting demand with and without the given technology, using a multistage stochastic programming model, whereas [30] uses a decision tree approach and production cost model. The system flexibility in [81] is addressed via the determination whether sufficient reserves are available and generator flexibility is considered over a generation pattern of a typical week. Generation technologies are aggregated by technology and commitment dates.

[82] develops a planning metric for flexibility offered by conventional power plants to meet net load changes. The Effective Ramping Capability (ERC) calculates the contribution of a single unit for the system to ramp in a given direction and time frame, thus calculating the flexibility available to the system operator via this unit. The calculation is made by comparing the system ERC with and without the assessed unit, both systems having the same Inadequate Ramp Resource Probability (IRRP), a metric for system reliability. The ERC is based on the forced outage rate and Ramping Availability Rate (RAR), measuring the probability that a unit will be able to offer its maximum ramp rate at any time. This metric is based on historical dispatch data.

[83] assess the value of a thermal power plant's operational flexibility improvements based on the unit's dispatch in response to market signals. [84] performs the same assessment but optimizes the whole system unit commitment in response to a demand profile. The dispatching algorithm in [83] relies on a comparison of price spreads and the cycling capability of the plant. The power plant's economic dispatch is not fully optimized, as the number, cost, and quality of start-up events are defined exogenously. The authors of

[85] use a dispatch model to assess the cost-effectiveness of storage technologies in a given scenario. The deterministic model optimizes the hourly dispatch to maximize the operator's profit over a year, and a net present value approach is used to obtain the value of an exogenously defined investment. The assessment in [83] includes the intraday market; it is, however, unclear whether the intraday price signal is used as the primary dispatching signal or as an arbitrage opportunity to correct the day-ahead contracted positions. The assessments in [84] and [85] include frequency control, the first one using exogenously defined control needs and the second one using the control remuneration. None of the studies reviewed here include both the intraday and frequency control market.

Table 2 Literature review of flexibility metrics

Metric name	Measure of:	Parameter	Reference
Insufficient ramping resource expectation (IRRE) metric Derived from Loss of load expectation (LOLE), a system reliability metric	Scheduled and realizable flexibility	Variability of the net load, using a probabilistic approach.	[57, 69]
Periods of flexibility deficit (PFD)	Periods of flexibility deficit, Scheduled and realizable flexibility	Net load ramps	[58, 69]
Expected unserved ramping (EUR), derived from expected unserved energy metric in capacity adequacy planning	Magnitude of the flexibility deficit, Scheduled and realizable flexibility	Net load ramps	[58]
Lack of ramp probability (LORP), derived from Loss of load probability (LOLP), a system reliability metric	Risk of ramp capacity shortage related to a dispatch decision	Minimum and maximum rated output, ramp-up and ramp-down rate, the current dispatch level of each generation resource and net load	[59]
Ramping Availability Rate (RAR)	Probability that a unit will be able to offer its maximum ramp rate at any time	Based on historical dispatch data	[82]
Effective Ramping Capability (ERC)	Contribution of a single unit for the system to ramp in a given direction and time frame	Forced outage rate and Ramping Availability Rate (RAR)	[82]

Metric name	Measure of:	Parameter	Reference
Flexibility trinity	Technical capability of individual units to deviate from their planned schedule	Power, energy, and ramp-rate	[16, 34] [38, 39]
Ramping availability	Upward and downward ramping available to the system within an uncertainty range.	Time, uncertainty, action, cost	[43]
Technical (Economical) uncertainty scenarios flexibility index (TUSFI and TE-TUSFI)	Transmission Network adequacy	Load flow distribution	[18, 19]
Transmission Network Adaptation	Adaptation time and cost of the transmission network	Load flow and investment costs for standard deviation of the additional cost for a plan to respond to an uncertain scenario; Expected Energy Not Supplied (EENS).	[66], [17] [67]
Transmission expansion flexibility (TEF)	Transmission Network adequacy	Expected unserved demand	[20]
Exportable flexibility	Transmission Network adequacy between TSOs	The reserve that can be exported via tie-lines	[68]

2.2 Specification of the solution approach

2.2.1 Static and dynamic methods

The various methods found in the literature for assessing operational flexibility could be placed into three categories: static, semi-static and dynamic. Static methods do not account for operational flexibility time scales; LCOE calculations or ordered load curves fall into that category [75]. The methods which use the power plant operation as a starting point for the assessment of its capability to deviate from this given trajectory as seen in [29], [16], [38] and [35] can be ordered in the second category. The differences in operation due to power plant upgrades are then only assessed with respect to this deviation. An assessment based on the full operation profile would be considered as a dynamic method.

To illustrate the limitations of static methods, one must consider assessing the value of a start-up cost reduction. Power plant start-ups are very costly regarding fuel as well as lifetime consumption; the expenses grow with the duration for which the critical power plant components (like thick-walled boiler parts or turbine rotor) have been cooling down. A straightforward method to assess a retrofit which reduces the start-up costs would be to set a given number of start-ups per year and to multiply this by the start-up cost reduction. This method would not take into account that with reduced costs, start-ups may come into competition with part-load events and thus increase in number and maybe over-compensate the costs due to degraded efficiency at part-load. Comparing two power plants with different start-up characteristics (be it costs, or duration) regarding operational flexibility by setting the number of events exogenously introduces a bias. The upgrades will not only impact the costs or duration of start-up events but also impact the number of start-up events. A further bias is introduced when assuming constant start-up costs. Since electricity is already produced during starts, it can also be sold, and thus the start-up net costs depend on the electricity price (but also fuel costs). An exemplary calculation shows the net costs (fuel costs minus sold electricity) in dependence on electricity prices, see Figure 2. Real coal-fired power plant start-ups have been positioned at different points in time in the German electricity market. The range of net costs for a given start-up procedure is unneglectable, as it reaches a factor of three in this case study.

To summarize, the energy market has a double dynamic, which influences the importance and costs of transient power plant operations. This dynamic is composed of (i) a dynamic which drives the actual costs and profits of flexible operation (start-up costs) and (ii) a dynamic which drives the number of cycling events within a given period (number of start-up events). Not accounting for these dynamics can substantially modify the results of the operational flexibility assessment. A further dynamic aspect of the considered problem is introduced by the timelines and time steps of the markets in which power plants are dispatched. The day-ahead, intraday and frequency control markets, as introduced in 0, differ by their timeframes, and these dynamic interactions should also be accounted for.

2.2.2 Time representation

In light of subsection 2.2.1, solving the problems introduced in section 1.4 requires a dynamic method. This subsection examines in detail the requirements of such a method regarding time representation.

Answering the research question requires a high time resolution. The high time resolution is necessary due to the high resolution of the markets themselves, like the hourly day-ahead markets, fifteen minutes intraday markets or the thirty seconds activation of the primary frequency control in Europe. Less straightforward, the time resolution of the assessment method has an impact on cycling costs, as demonstrated in [86–88]. [88] uses the energy system model PLEXOS to assess the impact of the time resolution on generation and start-up costs. Using an hourly time discretization instead of 5 minutes time resolution leads to a cost overestimation of 1 %, impacting the plant dispatch, start-up events, power flows and renewables curtailment. [86], also using PLEXOS but with regard to storage, demonstrates that time resolution impacts the value assessment of storage technologies. The authors of [87] show that shorter time resolutions reduce the impact of renewables' variability on the system balance energy.

Answering the research questions also requires long-term assessments. The time horizon needs to be long enough to include the seasonal patterns as well as different meteorological events, which are of importance in a system with a high share of variable renewable energy sources based on wind and sun. Operational flexibility impacts the unit condition, so that long time horizons are also required from that perspective.

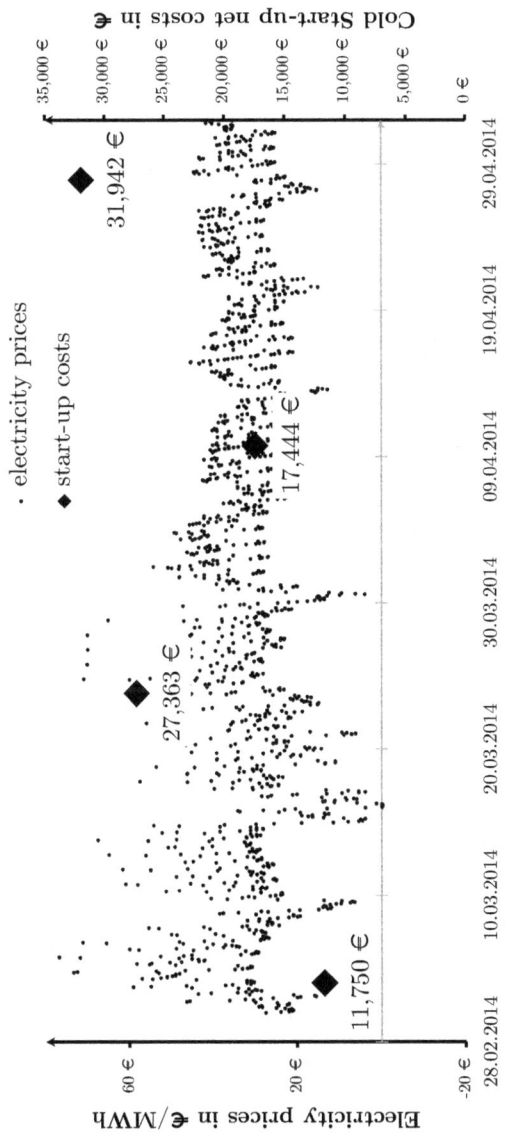

Figure 2 Net cost of a start-up in the German 2014 day-ahead electricity market as a function of time and market outcomes

The aspects of lifetime consumption and modified maintenance strategies are of special interest to the power plant operator. The overall operational strategies rely upon such considerations. Additionally, for the targeted assessment in a liberalized market, a key figure is the return on investment (ROI). Currently, European utilities are reluctant to invest in retrofits with ROIs greater than three years. This time horizon might appear as short-run from a planning perspective, but given the high time resolution required for the targeted research problem, time horizons of some years are qualified as long-term in our context.

Combining a long time horizon with a high time resolution is similar to combining planning and operation, which are usually separated in the literature [89]. When assessing operational flexibility, the burden is further increased by the necessity of combining different time scales and horizons [38, 87], as required when dispatching the plants in various interacting markets.

2.3 Research gap and contributions

The previous section 2.2.1 demonstrated the importance of dynamic methods for the research topic. However, most of the proposed methodologies screened in section 2.1 rely on static or semi-static methods. Furthermore, the power plant's operational flexibility is not limited to its technical capability, but on the incentives to make use of it in the power plant's liberalized market context. Therefore, the author proposes to make use of the power plant's dispatch as a flexible response to the signals set by its market environment, similar to the ideas raised in [83, 85]. In [83], the power plant dispatch is not fully optimized, as the number, costs, and quality of the starts are set exogenously. As the calculation in subsection 2.2.1 has shown, this does not suit our specification. Storage technologies are addressed in [85], whereas the work at hand targets conventional power plants.

In the literature, finding the power plant's dispatch is referred to as the unit commitment and economic dispatch problem. In [90], the idea of "using unit commitment models to value new products" is rated as a "very important topic requiring fundamental research". In this work's context, new products refer to power plant flexibilization. The work at hand contributes to this topic by quantitatively solving the long-term high-resolution unit commitment and economic dispatch problem for the valuation of power plant's operational flexibility. A unit commitment and economic dispatch problem suited for the research problem will be proposed.

As identified in the last paragraph of subsection 2.1.2.3, the impact of the interaction of the day-ahead, intraday and frequency control markets (markets identified in section 1.3) on operational flexibility has not been addressed previously in the literature. To answer this research question, the aforementioned model will be enhanced by these markets. This contribution assesses the value of flexibility when the intraday market is used as an arbitrage opportunity and capacity is sold on the markets for frequency control. The state-of-the-art opportunity cost approach to the frequency control capacity reservation will be improved to account for operational flexibility. As the literature review further shows, it is unclear how much detail a suited power plant description should entail. The work at hand thus contributes by proposing a parametric power plant model addressing the operational flexibility's relevant issues. A quantification of the influence of the chosen power plant model and problem formulation on the value of operational flexibility will be proposed to close these gaps.

The retroactive observation of the electricity market is used to answer the research questions. Scenario definition would be an alternative approach, with the drawback of making the results dependent on the scenario postulate. The European countries offer a broad range of renewables shares, so that the retroactive observation of the increasing shares with time, but also different penetration levels among countries, offers a solid basis for answering the research questions. The investigation of the market design and how the merchant dispatchable power plants react to the market in their operation strategies is the primary practical relevance of this work. The developed methods allow for a quantitative assessment of conventional power plants' operational flexibility. A set of dispatchable technologies will be compared in terms of operational flexibility in various market environments (day-ahead, frequency control, intraday, and combinations).

3 Unit commitment for product valuation

3.1 A map of the unit commitment world

3.2 Best suited unit commitment problem for the assessment of individual power plant's operational flexibility

3.3 Planning and scheduling, a vast world of approaches
3.3.1 Time-discrete approaches
3.3.2 Artificial intelligence: Temporal planning
3.3.3 Discrete Event systems: Automata theory

This section is dedicated to mapping the unit commitment field, thus helping the subsequent selection of a suited unit commitment problem for the research questions. Different approaches are available for solving this problem so that a review of their characteristics is performed.

3.1 A map of the unit commitment world

The unit commitment (UC) problem is defined in [91] as the problem of finding a commitment schedule minimizing the system costs or maximizing the player's profit for a set of generating plants over a given time horizon under a set of constraints. This problem is faced by generating utilities and system operators but also by supra-national, national and regional policymakers. In a liberalized context, the first one aims at operating and maintaining its fleets with profit maximization, whereas the second aims at operating the system from a welfare maximizing perspective, with a mind to system constraints. Policymakers can make use of this problem to analyze the effects of their policies and thus support decision-making. The unit commitment includes various facets and, depending on its purpose, might have very different formulations. The unit commitment can be complemented by the economic dispatch problem to derive the production level when power plants are committed [92]. If the network is accounted for, the problem is referred to as the optimal power flow problem [92]. The problem formulation consists of (and differs by) one or more objective functions and the set of constraints the market and technology imposes. With the liberalization, the European energy systems have seen a change from centrally dispatched systems to self-dispatch, and thus the unit commitment problem formulation has seen a major change of the objective function from cost minimization to profit maximization [93]. The drive toward more environmental-friendly systems is reflected by emission-constrained problem formulations [93], which limit the number of pollutants and price them. Further essential constraints for the system operation are the network ones. The network can be modeled with various degrees of complexity ranging from a "copper plate" assumption to non-linear and non-convex AC models [92]. With increasing shares of intermittent energy infeed, the system security issue is more significant, and services like frequency and voltage control are included in the problem formulation. The energy system electrification is facilitated by a more elastic demand, which also requires a modification of the problem formulation. The introduction of storage and vehicle-to-grid leads to moving borders between the supply and demand side, as these operate as well as sinks as generators. The problem formulations also vary in how uncertainty is represented. Purely deterministic problems ignore stochasticity, whereas stochastic problems include it in the problem formulation. Uncertainty can be accounted for in deterministic problem formulations by running the problem for multiple scenarios.

3.2 Best suited unit commitment problem for the assessment of individual power plant's operational flexibility

In Europe, most of the energy systems are organized as self-dispatch, whereas central dispatch is only in place in a few countries such as Greece, Ireland, Italy and Poland, see [47]. The main difference between both systems is the system operator's responsibility. Whereas in the centrally dispatched system the system operator collects all bids and defines the system operation and prices centrally, the self-dispatch system is organized in exchanges and over the counter transactions, which are communicated to the system operator along with the corresponding planned operation profiles. The system operator operates central control at this point and calculates costs for discrepancies after realization. Since the deregulation of most electricity markets, the electricity producers no longer operate the plant in agreement with a pre-defined load profile. Instead, they compete and strive to maximize their profit via optimal biddings. This situation can be described by the self-scheduling problem [94]. Thus, here it is proposed to solve the self-scheduling problem to perform the quantitative long-term high-resolution techno-economic assessment of operational flexibility at individual power plants. The purpose is to find the optimum dispatch of a price-taker merchant single power plant in response to the market price signals. In this sense, the problem is a unit commitment *and* economic dispatch problem. The deterministic perfect price foresight self-scheduling problem solved in this work is similar to the ones solved in [83, 85, 94, 95]. It relies on the assumptions that there is perfect competition as it is assumed that the dispatch decision for a single power plant does not affect the market prices [96]. This assumption is also referred to as infinite market assumption or price-taker approach. Bidding is not addressed. It has been demonstrated in [97] that under perfect competition, the generators should bid their marginal operation costs [96]. For readability, the deterministic self-scheduling problem of a merchant power plant from a price-taker perspective will be referred to as the self-scheduling problem in the following.

3.3 Planning and scheduling, a vast world of approaches

Formulating and solving the unit commitment and economic dispatch problem is a widely researched topic and extensive surveys on solving techniques can be found in [98], [91] and [93]. The world of scheduling power plants is vast by itself but is even more extensive if we consider the scheduling applied to other domains like the industrial scheduling of machines for production processes. This chapter, therefore, provides a review of the standard practices for power plant scheduling but also less conventional approaches coming from other disciplines.

3.3.1 Time-discrete approaches

Mathematical optimization means minimizing a function within an allowed set. The optimization of the unit commitment problem is time dependent. The problem formulation thus needs to define a time representation, which will influence the decision variable vector and the problem constraints. Depending on the use, the unit commitment literature makes use of a chronological time representation with the uniform discretization of the time horizon or aggregates time using commonalities in supply and demand structure (as seasons or week-days). This aggregation is often used when the problem includes long time horizons and a considerable amount of variables. This approach does not fit the specification of high time resolution required for our purpose. The amount of decision variables and constraints with constant time discretization is directly proportional to the number of time steps in the model [99]. A highly resolved time representation thus leads to a higher computational burden. The optimization problems faced in power systems further include non-linearity [100] and a mix of discrete and continuous processes [101]. The resulting problem would thus be a mixed-integer nonlinear programming problem (MINLP). Problems can be simplified to a linear problem (LP), and thus make use of effective solving techniques like the simplex method and interior point methods [100]. Problem simplifications are, however, made at the cost of the solution quality.

Mixed-integer programming (MIP) has become the approach of choice for solving the unit commitment problem. This has resulted from the drastic improvements in MIP problem solvers in the past years [102]. Research has been done and still works on the better formulation of the problem for lower solving duration [103], [104], [105]. However, most of the proposed problems are of moderate size and make use of simplistic (in the light of operational flexibility) linear models. The MIP formulation applied to generators with higher modeling precision results in possibly nonlinear power plant models and long computation time [106].

The burden related to the comprehensive enumeration of all possible solutions to these combinatorial problems has first been solved via the introduction of priority lists [107], based on rationales or guidelines such as the unit's incremental operation costs, similar to a merit order approach. Dynamic programming has been applied to the field of unit commitment for a long time, with applications already published in 1966 [108]. Dynamic programming is a search procedure in the space of possible unit commitment combinations, deploying the whole decision tree associated to the problem stages (typically hourly time steps) and states (combination of unit commitments). The main drawback of this method is its dimensionality. Methods have thus been developed to reduce the search space, see [98]. The use of dynamic programming is characterized by a recursive formulation and the definition of sub-problems [98].

A further search procedure has been introduced with the branch and bound approach [109], which finds near-optimal solutions in the space limited by a lower bound to the optimal solution. To find the lower bound to the optimal solution, the dual optimization problem needs to be solved [98]. This leads to the family of decomposition approaches, among which the Lagrangian relaxation [110], widely used and accepted in the community [98]. This method breaks down the problem into a master problem and its sub-problems. The dual problem, which links the master problem to the sub-problems via Lagrange multipliers, is more comfortable to solve. The Benders decomposition divides the problem into the economic dispatch problem and an integer unit commitment problem, thus solving the linear problem and searching for integer solutions. The literature offers a large variety of approach combinations like linear programming with dynamic programming [111] among many others.

All these methods intend to overcome the solution space size, related to the uniform discretization of time (the time sampling). Moreover, the discretization of time represents a problem approximation, which might lead to lower quality solutions. The following section's focus is thus on other time representations. A review of the time discrete and time continuous approaches to the problem of process scheduling is performed in [112]. The authors of [112] note that the characteristic of time-continuous approaches is the introduction of events or variable time intervals, which is also common to the approaches selected in the next sections. These less conventional approaches to the unit commitment and economic dispatch problem are further based on the fact that power plant operation is a controlled process with a finite set of actions and states.

3.3.2 Artificial intelligence: Temporal planning

Artificial intelligence based methods to solve the unit commitment and economic dispatch problem have been proposed and are reviewed in, e.g. [93] and [100]. Attempts have been proposed by [113], [99] and [114] to apply temporal planning to power plant scheduling. Temporal planning is a branch of automated planning and scheduling, itself a branch of artificial intelligence. Temporal planners do not require time discretization [114]. The primary adaptation of planning for unit commitment problems lies in the delivery of a best possible schedule (not only a feasible schedule but also optimum) and then in the representation of state-dependent action costs. The work of [114] showed that this approach is not competitive with MILP without a discrete time representation and requires proper heuristic guidance to deliver solutions of relatively good quality. Temporal planning is also close to the field of timed automata, which is discussed in the next session.

3.3.3 Discrete event systems: Automata theory

The continuous and discrete nature of processes in power systems has been analyzed from a control point of view in [101]. The author of [101] notes that Discrete events dynamic system (DEDS) might be a valid approach to solving unit commitment problems, as the time scale of the discrete commitment decisions is decoupled from the operational timescale. A discrete event system is defined in [115] as a dynamic system characterized by asynchronous occurrences of discrete, controlled as well as uncontrolled events. It is interesting to note that unit commitment and economic dispatch are also defined as two distinct problems and that the Bender's decomposition is based on the same distinction. This distinction might, however, be of less interest in a system with increasing variability, for which plants might be committed and de-committed on a daily basis.

Automata theory belongs to the DEDS theory and is defined in [116] as "the study of abstract computing devices, or "machines.""" and was originally used to model the human brain and define the tractability of problems for computing machines (Turing machine) [116]. The automata theory is characterized by the definition of state machines, with states and transitions, and the corresponding semantic defining the system behavior [117]. Timed automata extend this theory with a continuous time representation. Systems in which continuous and discrete dynamics coexist have also been studied from a control point of view using the automata theory, see, e.g. [118]. This approach sees a trajectory as a "sequence of continuous evolutions interleaved by discrete events" [118]. The finite state automata theory is used in the unit commitment field in combination with genetic algorithms for bidding strategies optimization in [90], chapter 9. Priced timed automata is used for fuel flow control at fossil-fired power plants in the Ph.D. thesis [119], for which model checking is used to find a feasible schedule with the property of lowest cost. In case of more complex systems involving concurrent processes, it has been proposed in [120] to make use of Petri nets, also belonging to DEDS theory, instead of automata.

4   Part conclusion

The first part of this work clarified the need for quantification of the value of operational flexibility at thermal power plants from a power plant operator perspective. The relevance of using dynamic approaches for assessing a power plant's operational flexibility has then been demonstrated. The importance of the temporal aspects, such as long time horizons and high time resolution, when it comes to operational flexibility has been discussed. A comprehensive literature review has been provided in light of these requirements, and the author, therefore, proposed to address the research topic by solving a unit commitment and economic dispatch problem to perform the quantitative long-term high-resolution techno-economic assessment of operational flexibility at individual dispatchable power plants. A review of the variety of unit commitment problems led to the selection of the deterministic single unit self-scheduling problem with a price-taker perspective and perfect price foresight. The next section then reviewed possible unit commitment and economic dispatch problem formulations and solving approaches, extended to the fields of planning, scheduling, and control. This broader overview indicated that methods differ primarily in how time is represented. Continuous time representations might be useful to overcome the approximation and computational burden introduced by uniform time discretization.

The next part of this work develops power plant models suited for the assessment of operational flexibility. The value of such a model is quantitatively worked out using state of the art mixed-integer programming formulations and off-the-shelf solvers. To overcome the limitations of the state-of-the-art approaches for formulating and solving unit commitment problems, the author further proposes a new approach, based on an alternative time representation. A further section deals with the identification of markets rewarding operational flexibility and extends the proposed model to the relevant markets. The benefit of including these markets in the analysis of operational flexibility assessments is then quantified. The applicability of all assumptions and simplifications made in this work is discussed in the last section.

Part B

Methods and concept development

1 On the importance of the power plant model

The authors of [121] quantitatively assess the effect of including more detail in the power plant description on the cycling behavior resulting from a system optimization. Similarly, this section aims to quantify the importance of the electricity generator description for an accurate plant-specific operational flexibility assessment from a plant operator perspective. Therefore, a quantitative assessment of the impact of the power plant description on the resulting optimized operation profile for (i) a state of the art power plant model (from the operations research literature) and for (ii) an improved generating plant is conducted. The aspects with the highest importance for the techno-economic analysis are first evaluated, before deriving a corresponding power plant description. Such a description is avoiding too much detail, in order to reduce the problem size, but includes all parameters influenced by operational flexibility. These include, among others, a detailed description of part-load operation from the perspective of increased lifetime consumption (see [122], [123]) and nonlinear efficiency decrease, leading to higher operating costs and specific CO_2 emissions. Numerical experiments are conducted to provide a quantitative sensitivity analysis, showing the influence of the power plant description on the results.

1.1 Flexibility parameters

The way power plants are represented is highly dependent on the use of the model. For a whole energy system analysis, power generating units are aggregated by type of fuel or technologies due to the high amount of data and the related computational burden. The power plant is defined via its load, how this load can vary and the corresponding costs. This very high-level description often aims at an economical, or techno-economical, system planning analysis. From a control point of view, the model aggregation might need to be reduced and include some power plant components (like the mills or heat exchangers), thus allowing for more insights in the technical processes at the plant. Other approaches aim at analyzing the impact of flexible operation on the steam cycle or material at critical components and use, e.g., fluid dynamics for very detailed technical studies. Depending on the aim of the analysis, it is thus crucial to reach a trade-off between model accuracy and computational tractability. For the analysis of current and future operation of conventional power plants in markets with growing shares of renewables, part-load and transient operations will increase. The resulting lifetime consumption and maintenance expenditures, the increased specific emissions, and reduced efficiencies are part of the techno-economic power plant model for flexibility assessment proposed in this work.

1.1.1 Nomenclature

1.1.1.1 Indices and sets

$t \in \Omega$ Instant time t in the time range Ω

$\Omega_i \in \Omega$ Time interval $\Omega_i = [t_i, t_{i+1}[$ with $i \in [0, N]$. The time range Ω is divided in N slices Ω_i, which are typically equal and correspond to the time step $\Delta t = t_{i+1} - t_i$ of the electricity market

$k \in K$ Transition counter k, characterized by the issuing time $t_k \in \Omega$

1.1.1.2 Spaces

$\mathcal{H}$ Space of functions $\Omega \to \mathbb{R}$

$\mathcal{H}_{x_j}$ Space of functions $\Omega \to [\check{x}_j, \hat{x}_j] \subset \mathbb{R}$, $[\check{x}_j, \hat{x}_j]$ the value range of the power plant parameter x_j.

1.1.1.3 Exogenous Parameters

$EP(t)$ Electricity price at time t in €/MWh

$FP(t)$	Fuel price including CO_2 emission costs at time t in $€/MWh_{th}$
$\hat{P}(\check{P})$	Maximum (resp. minimum) power output of power plant in MW_{el}
RU (RD)	Ramp-up (resp. ramp-down) rate of power plant in MW_{el}/h
TU (TD)	Minimum up-time (resp. down time) in h
f_0	Fuel consumption at idle, in MW_{th}
f_1	Fuel consumption at nominal load, in MW_{th}
f_{min}	Fuel consumption at minimum load, in MW_{th}
cm_0	Maintenance costs, static continuous component during standstill in $€/h$
cm_h	Maintenance costs, static component during plant availability in $€/h$
cm_1	Maintenance costs, load proportional component in $€/MWh_{el}$
ce_k	Transition specific maintenance costs in $€$
f_k	Transition specific integrated dynamic fuel consumption in MWh_{th}

1.1.1.4 Endogenous Parameters

$p(t)$	Instant power plant output at time t, in MW_{el}
$f(t)$	Instant power plant fuel consumption at time t, in MW_{th}
$c_m(t)$	Maintenance costs as function of time t in $€/h$
$u(t)$	Binary commitment variable which equals 1 if the power plant is above its minimum technical load at time t, otherwise 0
$\delta(t - t_k)$	Dirac function which equals 1 if $t = t_k$, 0 else
$y(t)$	Load proportional component of the objective function, as function of time t in $€/MWh$
$y_0(t)$	Load independent component of the objective function, as function of time t in $€/h$

1.1.2 Power plant cycling

The cycling of power plants is clustered into different types of processes, with the highest level grouping distinguishing between start-up, part-load cycling, and shutdown. These will be referred to as "transitions" in the following sections. Start-ups can then be further clustered along the temperature of critical components such as the turbine rotor. The most common types are those of cold, warm and hot starts, which, by extension of this definition, sometimes stems two additional intermediary temperature ranges. Knowing the cooling process of these critical components, the start-up types can also be distinguished with the duration for which the components have been cooling down. Temperatures and cooling durations are reviewed in [14], as these depend on the plant characteristics. Start-ups are here defined as ranging from the first start-up fuel ignition toward the load reaching its minimum level. Start-up fuel (or backup fuel or ignition fuel) is a supplemental firing of oil or gas to stabilize the flame in a coal-fired boiler. The ignition fuel is also required to start-up the power plant and is more expensive than coal. The minimum load level is defined as the lowest possible stable and environmental compliant load production. Part-load cycling is the action of ramping up and down between the minimum load level and the full load level. These part-load transitions can be distinguished along the ramp speed and maintenance costs. The power plant model introduced in this work distinguishes between the ramp-rates after each start-up type, before shutdown and the load following ones. Ramp-rates refer to the load difference between two steady states divided by the time required for the power plant to change the load from the first level to the second one. In this work's context, the ramp-rate is thus considered constant over the whole minimum load to full load range.

1.1.3 Fuel consumption and efficiency

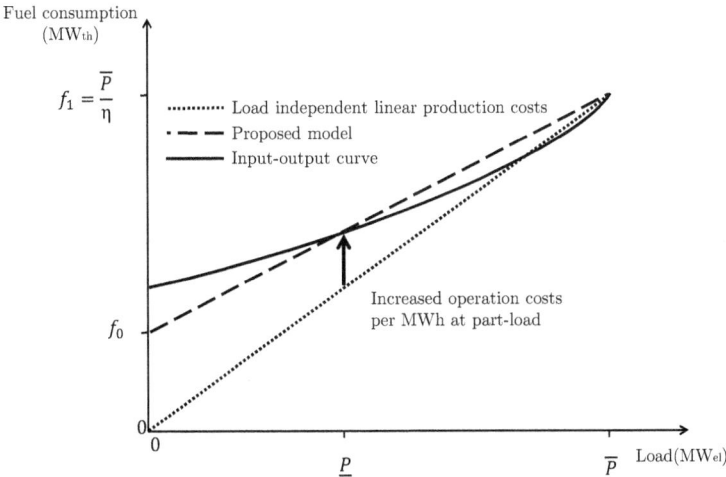

Figure 3 Model describing the evolution of the fuel consumption at part-load

When the power plant is operated at part-load, its efficiency is decreasing. The fuel consumption as a function of the unit's output is a non-linear relationship, which is often approximated by a quadratic function [100]. For linearity, this function can be further approximated by a (stepwise) linear function. Approaches assuming load independent variable production costs like in [95] (corresponding to a directly proportional fuel consumption illustrated in Figure 3) ignore the efficiency decrease at part-load. To overcome this caveat, but with regard to the problem tractability, the fuel consumption at steady state, once the plant is synchronized to the grid, can be written approximately by a linear load dependent equation. The thermal fuel consumption varies between the strictly positive fuel consumption at idle and its full load value. This relationship describes a steady state pictured in Figure 3. The model proposed in this work explicitly accounts for the fuel function in the problem formulation. In addition to the efficiency evolution with load, this further allows accounting for the fuel price variations, particularly relevant when performing long-term assessments. By every time t_k dynamic load changes (transitions, like start-ups) occur, a dynamic element f_k is added to the static linear fuel consumption model. For part-load operations which consist of the combination of a ramp-down and ramp-up phase, one may wish to simplify the problem and thus assume that these dynamic fuel consumptions compensate each other. The fuel saved during the ramp-down phase thanks to the stored heat will be approximately compensated by the fuel required to ramp-up the plant. The fuel model resumes to the following equation (1):

$$f(p(t),t) = \sum_{k \in K} f_k \cdot \delta(t - t_k) + f_0 + p(t) \cdot \frac{1}{P} \cdot (f_1 - f_0) \quad \forall\, p(t), t \qquad (1)$$

1.1.4 Greenhouse gas emission costs and limitations

To reduce greenhouse gas emissions, some countries price or cap them. Pricing mechanisms are distinguished between carbon taxes and capping and trading systems. The second mechanism has the advantage of requiring the definition of a carbon budget, and thus to drive the prices on a demand and supply basis. The example of the European trading system, however, demonstrates how the inappropriate dimensioning of this budget can make the system ineffective. Some countries thus introduced national carbon prices, see Part C section 2.3.3. Emissions are thus a crucial topic for the economics of power plants. In a European context, the European allowance system (EU ETS) for CO_2 pricing has to be taken into account in the calculation of marginal operation costs, even if the current price level of the EU ETS is shallow. Other emissions like SO_2 and NOx as well as particulates are either priced or at least regulated. The relevant European policies are the integrated pollution prevention and control directive (IPPC), the National Emissions Ceiling Directive (NECD, 2001/81/EC) and Large Combustion Plants Directive (LCPD). This topic is more significant at part-load and minimum load operation: since the efficiency is decreased, there are more specific greenhouse gas emissions. Furthermore, the decreased operating temperatures are problematic for catalytic devices for the capture of NOx. Thus the presented fuel model does not only allow keeping track of the fuel consumption but also of the emissions via the use of emission factors. Attention must be paid to the fact that flexible operation requires using the fuel specific emission factor (means in mass per thermal energy) and not the electric emission factor (which would not account for the increased emissions at part-load, as in [83]).

1.1.5 Lifetime consumption and maintenance costs

The lifetime of a component is the duration for which it can be operated safely. Limitation in lifetime applies to components exposed to temperatures above 400 °C [122]. The creep damage mechanism occurs during steady-state operation at static stresses. Additionally, the material properties decline with high working temperatures, so that operation at a higher temperature has a higher creep impact. With increasing cyclic operations, low cycle fatigue (LCF) plays a greater role. It is caused by high thermal stresses in the components, resulting from temperature gradients during heating and cooling phases of thick-walled components. Both of these mechanisms influence the maintenance costs of the power plant, with an increasing share of LCF. The maintenance costs need to be broken down into different categories: maintenance costs which continuously add up during the course of the year (static continuous component), maintenance costs which add up depending on the operated load (load proportional and static component) and transition-based maintenance costs, which add up whenever a transition is executed. The continuous static component adds up over the whole plant lifetime, independent of its operation and comprises aging of components such as creep, consumables like ammonium for flue gas treatment as well as crew costs. In the operation regime above minimum technical load, a load proportional part comprises aging of components such as creep damage due to pressure or centrifugal forces. The static component is increased due to higher crew and consumables costs. The maintenance costs are illustrated in Figure 4. During transitions, a specific dynamic maintenance cost component c_k is added. In the study [124], North American power plants have been analyzed to determine the effect of cycling on their operation and maintenance costs. One effect which has been found there but is not accounted for here is the moderate effect on heat rates (see page 35 [124]). However, it can easily be taken into account by sliding the linear fuel function upwards. The maintenance cost model resumes to the following equations (2) and (3). The proposed formulation requires the definition of the function (3) for keeping track of the power plant state.

$$cm(t) = cm_0 + cm_h \cdot u(t) + cm_1 \cdot p(t) \cdot u(t)$$

$$+ \sum_k ce_k \cdot \delta(t - t_k) \tag{2}$$

$$
\begin{aligned}
u : \; & \Omega \to \mathbb{R} \\
u(t) = 0 \quad & \forall p(t) < \underline{P} \\
u(t) = 1 \quad & \forall p(t) \geq \underline{P}
\end{aligned}
\tag{3}
$$

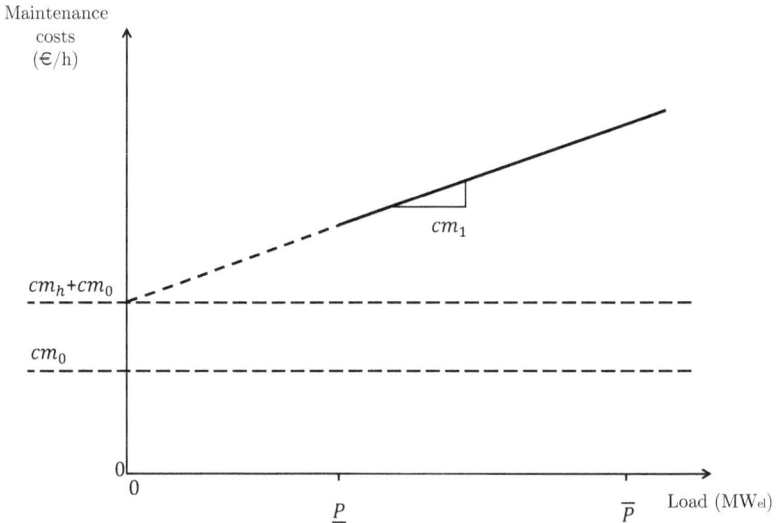

Figure 4 Model for the evolution of the maintenance costs over the operation regime

1.2 The optimization problem formulation

The self-scheduling problem formulation is a combination of an objective function and the problem constraints. The power plant description introduced in the previous section is used to formulate the problem's objective function and implemented in a mixed integer programming (MIP) formulation of the problem.

1.2.1 Self-scheduling objective function

The self-scheduling problem's objective function is the power plant operator's profit function, which integrates the difference between the sold electricity and the operation and maintenance costs as in equation (4). The parameter $EP \in \mathcal{H}$ is the electricity price, $FP \in \mathcal{H}$ is the fuel cost, the variable $p \in \mathcal{H}_p$ is the electrical power, $f \in \mathcal{H}_f$ is the fuel requirement and $c_m \in \mathcal{H}_{c_m}$ is the maintenance cost function. The optimization has to find the optimum by a variation of the functions p, f and c_m spanned by the combined functional space $\mathcal{H}_p \otimes \mathcal{H}_f \otimes \mathcal{H}_{c_m}$, while the electricity price and the fuel costs are determined exogenously by the market. The main assumption for problem tractability is to reduce the solution space from the functional space $\mathcal{H}_p \otimes \mathcal{H}_f \otimes \mathcal{H}_{c_m}$ to the space $\mathcal{H}_p$ via the assumption that the fuel and maintenance functions are time independent functions of the power production p. Using the fuel and maintenance functions introduced previously, the unit commitment problem has the profit maximizing objective function of equation (5).

$$\max_{\mathcal{H}_p \otimes \mathcal{H}_f \otimes \mathcal{H}_{c_m}} \int_\Omega [\, EP(t) \cdot p(t) - FP(t) \cdot f(t) \; - c_m(t)]dt \tag{4}$$

$$\max_{\mathcal{H}_p} \int_\Omega \left[\left(y(t) - cm_1 \cdot u(t) \right) \cdot p(t) + y_0 - cm_h \cdot u(t) - \sum_{k \in L} f_k(t) \cdot \delta(t - t_k) \right. $$
$$\left. \cdot FP(t) - \sum_{k \in L} ce_k \cdot \delta(t - t_k) \right] dt \tag{5}$$

$$y : \Omega \to \mathbb{R}$$
$$y(t) = EP(t) - (f_1 - f_0) \cdot \frac{1}{\overline{P}} \cdot FP(t) \tag{6}$$

$$y_0 : \Omega \to \mathbb{R}$$
$$y_0(t) = -(f_0 \cdot FP(t) + cm_0) \tag{7}$$

1.2.2 Uniform time-discrete problem formulation

The mixed integer nonlinear programming (MINLP) formulation of the self-scheduling problem (referred to as $TimeDiscreteNL$ in the following) is the combination of the time-discrete version of equations ((5)), (6) and(7) in the chosen time partition $\Omega = \bigcup_{i=0}^{N-1} \Omega_i$ and the problem constraints. The problem constraints are limited to power plant constraints, as in our case the network is not taken into account and that the power plant operation is market driven (and not demand driven). The controllable power plant constraints found in most problem formulations constrain the plant load within its minimum and maximum bounds, limit the plant load from one time step to the next via the plant ramp-rates, prevent the plant from starting or shutting down if minimum up- and down-durations are not reached and link the start-up type to the time the power plant critical components have been cooling down. The focus of this work is not on the constraints formulation, so that the equations (1), (2), (3), (4), (5), (6), (7) and (8) developed in the model of [103] are used. This formulation has been chosen due to its tightness, accurate differentiation between power and energy and possibility to integrate as many transitions as required. The same variable denomination has been kept for ease of use. However, it should be noted that the power output p there designates the power above the minimum technical load. A different time partition can easily be introduced. The MILP problem defined in [103] will be used as reference and referred to as $TimeDiscreteL$ in the following.

1.3 Sensitivity of the optimal dispatch to the power plant description

Numerical experiments are conducted to assess the sensitivity of the optimal dispatch to the power plant description. Table 3 summarizes the different cases that have been defined with their naming. Two power plant models are distinguished: one containing the flexibility description as developed in section 1.1 and typical models from the literature which do not. The parameter cm_1, when different from zero, introduces a non-linearity.

Table 3 Test case definition and naming for the quantification of the power plant model influence on the dispatch optimization

Case identification	Problem approach	Power plant description
TimeDiscreteL	MILP	Without flexibility parameters
TimeDiscreteNL	MINLP	With flexibility parameters

1.3.1 Assumptions for the state-of-the-art power plant description

The parameters of the proposed test cases are assigned values which can be found in Table 4 to Table 6. The values for the *TimeDiscreteL* power plant are chosen to depict the same technical capabilities as the reference power plant from [103]. The fuel consumption at idle is set to zero to obtain a constant efficiency evolution over the feasible load range, and the static maintenance costs are also set to zero in order to have no costs when the plant is not operating. The electricity price curve is to be found in Table 10. The fuel prices are constant in order to obtain constant production costs.

Table 4 Value of the economic test parameters for the *TimeDiscreteL* case

Economic power plant description					
FP $[\text{€/MWh}_{\text{th}}]$	cm_0 $[\text{€/h}]$	cm_1 $[\text{€/MWh}_{\text{el}}]$	cm_h $[\text{€/h}]$	$f_{dyn,SD}$ $[\text{MWh}_{\text{th}}]$	ce_{SD} $[\text{€}]$
22	0	0	200	9	222

Table 5 Value of the technological test parameters for the *TimeDiscreteL* case

Technologic power plant description					
f_0 [MWth]	η_b [-]	$\overline{P}$ [MW$_{\text{el}}$]	$\underline{P}$ [MW$_{\text{el}}$]	TU/TD [h]	RU/RD [MW$_{\text{el}}$/h]
0	0.4	378	150	3	76

Table 6 Value of the start-up test parameters for the *TimeDiscreteL* case

Start-up ramping information					
Start-up type	1	2	3	4	5
$f_{dyn,SU}[\text{MWh}_{\text{th}}]$	5	9	14	19	23
ce_{SU} [€]	106	230	328	422	535

The start-up curves have an hourly linear evolution of the load between 0 MW and $\underline{P}$ MW

1.3.2 Assumptions for the improved power plant description

The parameter changed to account for the flexibility parameters presented in section 1.1 are to be found in Table 7. The fuel consumption f_0 at idle accounts for the efficiency drop at part-load. The chosen value indicates that at minimum load, the power plant efficiency is 0.39, instead of 0.4 at full load. The increased aging due to ramping is modelled using the parameter cm_1 while the costs incurring, even if the plant is not operated, are a combination of the parameters cm_0 and f_0. In addition, the fuel prices are not held constant any more. The chosen function shows a drop in fuel price corresponding to e.g. a new delivery price.

Table 7 Value of the modified techno-economic test parameters for the *TimeDiscreteNL* case

Parameters for the sensitivity analysis

FP [€/MWh$_{th}$]	cm_0 [€/h]	cm_1 [€/MWh$_{el}$]	f_0 [MW$_{th}$]
Drop from 22 to 20 €/MWh at the 47th hour	65	1	15.12

1.3.3 Simulation parameters

The MILP tests have been carried out using Matlab R2014b and CPLEX 12.1. The MIP problems are solved to 10^{-6} of relative optimality tolerance. The MINLP tests have been carried out using Matlab R2014b and the solver SCIP [125, 126]. All tests have been carried out on an Intel® Core ™ i5 CPU M520 with 2.4 GHz and 4 GB of RAM.

1.3.4 Quantification of the flexibility parameters importance

The sensitivity analysis assessing the proposed power plant model's effect on the resulting optimal dispatch is to be found in Table 8. The $TimeDiscreteNL$ case implements the power plant model described in section 1.1. It is compared to a reference case, which is the $TimeDiscreteL$ case with the reference power plant description. Each result shows the change in profit due to the combined modification of the parameters indicated horizontally and vertically. In row 2, column 1, the value shows that using the non-constant fuel prices and the non-zero fuel consumption at idle (reduced efficiency at part-load) results in a profit 8 % lower than the profit obtained without these parameters. The values resulting from this analysis quantitatively evaluate the importance of the generator description and allow to conclude that it influences the operational flexibility valuation. All the aspects introduced in the power plant model cannot be neglected without having a significant influence on the results. The parameter with the highest impact, in this case, is the part-load efficiency decrease, despite the modest efficiency decrease assumption. The results of the one-at-a-time sensitivity analysis are further depicted in Figure 5. The figures show the profit reached when variating the enhanced power plant description in the range defined in Table 7. The y-axis scales, with variations from a factor 10 to 100 further illustrate the importance of the part-load efficiency degradation.

Table 8 Change in optimum profit due to enhanced power plant description. Each result shows the relative increase in profit due to the combined modification of the parameters indicated horizontally and vertically.

Modified parameter	Variable fuel costs FP	Non-zero f_0	Non-zero cm_0	Load-dependent maintenance costs cm_1
Variable fuel costs FP	Test1: +32 %	-	-	-
Non-zero f_0	Test2: -7.9 %	Test3: -43.8 %	-	-
Non-zero cm_0	Test4: +21.6 %	Test5: -54.3 %	Test6: -10.5 %	-
Load-dependent maintenance costs cm_1	Test7: +20.8 %	Test8: -52.5 %	Test9: -19.2 %	Test10: -8.7 %

Profit in €

f_0 in MW

FP in €/MWh$_{th}$

cm_0 in €/h

cm_1 in €/MWh

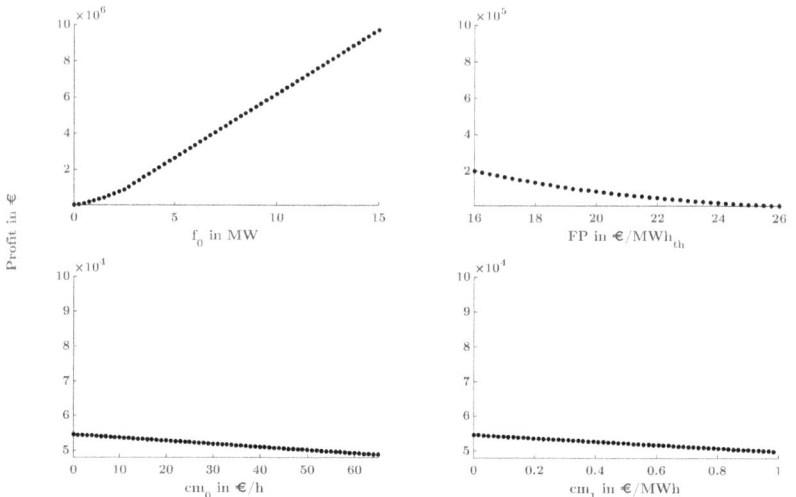

Figure 5 Optimum profit as a function of the enhanced power plant description's parameter values. Results of the one-at-a-time sensitivity analysis.

1.4 Section conclusion

The importance of the power plant description when performing the long-term high-resolution techno-economic assessment of operational flexibility has been demonstrated and quantified. The most critical aspects for flexible power plant operation have been described, and linear as well as non-linear models have been proposed, which can be used in state-of-the-art mixed integer programming (MIP) formulations of the problem (almost every energy system analysis model). The importance of the power plant model on the results further raises the question of the influence of the optimization problem formulation on the results. The review of the approaches to the unit commitment problem in the first part, section 3.3, has shown that the time discretization represents a problem approximation. The impact of this approximation is reduced when using a high-resolution discretization, which, however, increases the number of problem variables; a drawback, especially for longer time horizons and higher technical resolution. This might lead to more substantial, nonlinear problems. The next section thus introduces a problem specific formulation and solving heuristic for the self-scheduling problem from a price-taker perspective.

2 Event-based optimization

2.1 Theory

The continuous and discrete nature of processes in power systems has been analyzed from a control point of view in [101]. This mixed nature is one of the reasons why mixed integer programming techniques apply so well to solving unit commitment problems. The author of [101] however notes that Discrete events dynamic system (DEDS) might be a valid approach to solving these problems, as the time scale of the discrete commitment decisions is decoupled from the operational timescale. It is interesting to note that unit commitment and economic dispatch are also defined as two distinct problems. In the DEDS theory, trajectories are event-driven [101]. The automata theory, which belongs to the DEDS theory, defines finite state machines, which inspired the approach presented in the following. The event-based optimization technique proposed here solves the merchant power plant's self-scheduling problem from a price-taker perspective. Three features are combined: the first one is a generating unit evolution model in the form of a finite state machine sophisticated enough to include thermo-economical changes initiated by flexibility improvements but also efficient enough to determine the actual values of power plant internal parameters.

Secondly, a minimized optimization space is achieved by the introduction of a sorted list of events, a fundamental notion to avoid the use of a uniform time discretization. Moreover, a heuristic search approach which allows for efficient identification of optima is developed.

2.1.1 Nomenclature

2.1.1.1 Indices and sets

$t \in \Omega$ Instant time t in the time range Ω

$\Omega_i \in \Omega$ Time interval $\Omega_i = [t_i, t_{i+1}[$ with $i \in [0, N]$. The time range Ω is divided in N slices Ω_i, which are typically equal and correspond to the time step $\Delta t = t_{i+1} - t_i$ of the electricity market

$a, b \in \mathbb{S}$ States of the generating unit specific state machine, $\mathbb{S} = [1, ..., n_s] \in \mathbb{N}$

$a \to b \in \mathbb{T}$ Transition from state a to state b, in the set $\mathbb{T} \subset \mathbb{S}^2$ of transitions

$e_k \in \mathcal{E}$ Single events $e_k = (t_k, a \to b, z_k)$ with the event counter k, characterized by the issuing time $t_k \in \Omega$, the transition $a \to b$ and a control parameter z_k.

$\mathcal{E}$ Event space $\Omega \otimes \mathbb{T} \otimes \mathbb{R}$ of possible events

$E \in \mathcal{E}^K$ Sorted list $(e_1, ..., e_K)$ of $K \in \mathbb{N}$ events, with strictly increasing issue times $t_k + T_k \leq t_{k+1} \forall k < K$, and T_k as the duration of the transition issued by e_k.

2.1.1.2 Exogenous Parameters

$y \in \mathbb{R}^{m_y}$ Exogenous parameters with a dimensionality m_y. Examples are the electricity price, the fuel price and ambient air temperature

$EP(t)$ Electricity price at time t in €/MWh

$FP(t)$ Fuel price including CO_2 emission costs at time t in €/MWh$_{th}$

$\hat{P}(\check{P})$ Maximum (resp. minimum) power output of power plant in MW$_{el}$

$RU (RD)$ Ramp-up (resp. ramp-down) rate of power plant in MW$_{el}$/h

2.1.1.3 Endogenous Parameters

$x \in \mathbb{R}^{m_x}$ The parameters of the power plant with a dimensionality m_x. Examples are power, fuel consumption, and rotor temperatures.

$x_s \in \mathbb{R}^{m_s}$ The parameters of the power plant with a dimensionality m_s relative to state s.

$p(t)$ Instant power plant output at time t, in $\mathrm{MW_{el}}$, $p \in [0, \hat{P}]$

$f(t)$ Instant power plant fuel consumption at time t, in $\mathrm{MW_{th}}$

$c_m(t)$ Maintenance costs as function of time t in €/h

2.1.2 Power plant dispatch model

Generating plants are usually controlled by automated systems reacting and correcting on external conditions modifications. In most cases, closed-loop controls react continuously to such discrete changes [101].The idea developed in this section is similar to the approach in [118], which defines a trajectory (in our case the power plant evolution or dispatch) as a "sequence of continuous evolutions interleaved by discrete events."

The evolution of the power plant can be described via a differential equation of the form

$$\dot{x} = G(x, y, t) \qquad (8)$$

which defines the evolution of the power plant variables $x \in \mathbb{R}^m$ (e.g. turbine load or rotor temperature), as a function of plant variables x and of exogenous parameters y (such as air temperature or electricity prices) via a generating unit specific function G. An event is initiated when an intended control or unintended input changes value, which may be regarded as further exogenous parameters. Events are defined in [112] as a time instance in the continuous time domain, associated to continuous problem variables. The generating unit function for techno-economical calculations may be simplified by introducing the formalism of a state machine, which connects states a and b by intended transitions such as $a \to b$. The evolution of the state specific, internal parameters is defined by a state specific evolution operator with a similar form to (8). With a state machine, the generator status is defined by the status identification s and the value of the status specific internal parameters x_s. The status might depend on a much smaller number of internal parameters than the state independent variable x . For convenience, the state independent variable denomination x is used for x_s. This is also valid for the evolution of transition specific parameters, as they are controlled with a strictly deterministic behavior.

Once the generator has reached a state, it remains in this state until an event issues a transition. The number of possible events leading to a transition to another state is constrained via the state machine. An intended event $e_k = (t_k, a \to b, z_k)$ is characterized by the time t_k it is issued, the start state a and the target state b and some control parameter z_k. The evolution of the internal parameters x during the transition $l = a \to b$ depends on the event parameters as well as the initial value $x_k = x(t_k)$. An example is the internal parameter "rotor temperature", which value at the issue time (t_k) defines the resulting ramp-rates of a cold, warm or hot start.

The evolution of the generator (its dispatch) is entirely described by the state machine and the time ordered list of intended events $E = (e_1, ..., e_K) \in \mathcal{E}^K$, as the generator specific state machine defines states and transitions, described by the state evolution functions $G_s(x, y, t)$ and the transitions evolution tions $G_{e_k}(x, y, t)$. If, in addition, the boundary conditions of the generator are defined as well, then all values of the internal parameters x are known for the time interval Ω. The boundary conditions of a state machine contain the state and the value of the state specific internal parameters at time t_0. A list of intended events is finite as each transition has a non-zero finite duration time T_k determined by the event, whose sum over all events is bounded by the time interval Ω.

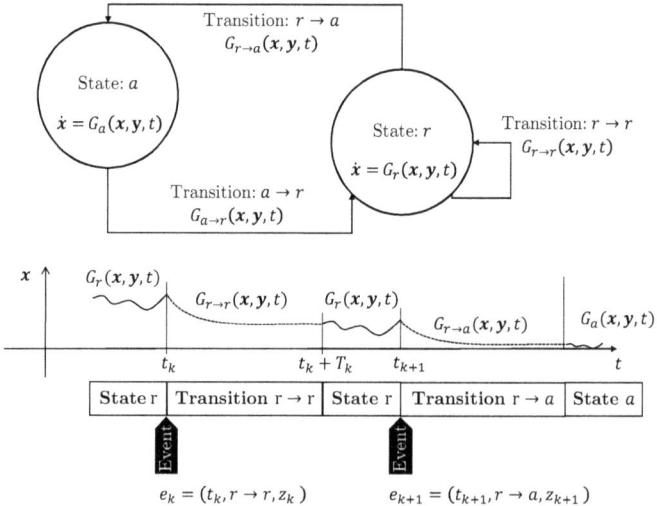

Figure 6 Top: generic state machine. Bottom: a generic development of the internal parameters as a function of time. Transitions are issued by events.

Figure 6 shows a generic state machine. The generating plant is either in the state "running" (r) or in the state "available" (a). Being available, the plant might start-up at any time. The start-up depends on the time elapsed since the last shutdown. It is conventional to differentiate between three to five start-up types, in dependence of the time the rotor has been cooling down. Cold, warm or hot starts are having different ramp-rates, maintenance costs and deterministic controlled behaviors. Start-ups are defined as the process ranging from the first support fuel ignition until the power plant's minimum stable load has been reached.

2.1.3 Event-space optimization

Any optimization of the problem's objective function requires updating the dispatch of the generator. Typical heuristics start from a first guess and then iteratively improve the solution. The initial solution could result from a random pattern or any further form of pattern generating this set of events. Here a pattern based on the functioning of the electricity market is however preferred. The evolution of the generator is known from the time ordered event list $E = (e_1, ..., e_K)$, with $e_k \in \mathcal{E} \; \forall k \leq K$ and the starting values. Accordingly, the objective optimization of equation (4) may be translated into equation (9) where the integral is fully determined for a fixed event list.

$$\max_{E \in \mathcal{E}^K} \int_\Omega [\, EP(t) \cdot p(t, E) - FP(t) \cdot f(t, E) \, - c_m(t, E)] dt \qquad (9)$$

Thus, the optimization process must-run through the following sequence: (i) define an event list, (ii) deduce the power, fuel usage, and the maintenance cost as a function of time (iii) integrate the values. The event list is modified, and the sequence runs again. The event list with better objective function value is kept, the other ones rejected. This process is illustrated in Figure 7 and explicated in the following.

2.1.3.1 Merit order ranking for initial event list guess

The marginal cost of the power plant may be used to retrieve a first event list. The algorithm starts with the determination of the operation and availability states which respectively correspond to an infra- and supra-marginal ranking in the merit order. If the electricity price is higher than the marginal cost, then the generator is operating. Otherwise, it is only available (clean dark or spark spread approach). The algorithm proceeds by detecting times when the electricity price crosses the marginal costs. Let k be the last added event. If the electricity price crosses downwards at time t, then a shutdown event is inserted to the event list at time t. If the price crosses upwards, then the time difference $t - t_k$ is used to determine the transition time T_{k+1} of a possible start-up ending at t. The duration $t - t_k$ corresponds to the cooling duration of the critical components and therefore determines which type of start-up can be operated. If $t > t_k + T_{k+1}$ this new start-up event is added to the event list. Else, the shutdown event k is cancelled. The algorithm continues until the end of the time range. The process results in an initial guess for the power plant dispatch profile.

2.1.3.2 Event optimization

The continuous parameters of an event are the issue time and the control parameter z. Both parameters may be optimized for each event to improve the income of the generator. However, events are coupled to direct neighboring events since start-up processes are affected by the time since the last shutdown, often even by a discontinuous relationship. Because of this, a combined optimization of both events is conducted. The considered time interval is divided in time slices, which are much shorter than the intervals defined by the electricity market. The size of the time slices will be referred to as the time step. It is not comparable to the time discretization of a MIP approach, as in the event-optimization case the time representation is not uniformly divided into the said time slices over the whole optimization time horizon but rather only around the event issue time.

2.1.3.3 Event-list optimization

The event list optimization underlies a search-tree. Its branches correspond to changes in the event list by combining, creating or suppressing events. The event modification rules are defined by the generator specific state machine. In the case of the generic state machine shown in Figure 6, a shutdown and startup sequences $(r \rightarrow a, a \rightarrow r)$ may either be replaced by a part-load $(r \rightarrow r, r \rightarrow r)$ operation or the event sequence may also be annihilated. Accordingly, the number of events is not persistent during the optimization process.

2.1.3.4 Optimization process

When referring to the optimization process, states do not designate the power plant states but the nodes of the search tree. The process starts with the definition of the first guess in the form of an event list, which can be found via the clean spread approach described above. The event list singleton $\{E_0\}$ is the initial state. The number of events K in $E_0 = (e_1, \ldots, e_K)$, with $e_k \in \mathcal{E} \; \forall k \leq K$ defines the stages or iterations of the algorithm. For each of these stages, the first task is to select the set of feasible states $\{I\}$ among the set of all possible states $\{J\}$. In order to do so, changes are made to the singleton $\{E_{k-1}\}$ defined at the previous stage. Each modification is an element of $\{J\}$. If the modification does not affect the events before e_k and is allowed by the state machine semantic, then it is an element I in $\{I\}$. After a change of the event list, a new event optimization must be conducted and the quality of the solution measured via the equation (9). Once this has been made for all elements of $\{I\}$, the event list I_{opt} in $\{I\}$ with highest quality (the highest profit) is selected and defines the new singleton $\{E_k = I_{opt}\}$. The same process iterates at the next stage, until the final stage K.

The phase space reduction is performed in numerous manners: first, the search tree is reduced using the state machine semantic. Then, at each iteration, the branches with lowest solution quality are fathomed to keep the local optimum identified at the said stage. The time partition of the approach, with a continuous time representation over the optimization horizon and a discrete time representation around the event initiation time t_k, also contributes to the phase space reduction. A further phase space reduction lies in the state machine itself, as it can be defined to allow transitions between minimum load and nominal load without considering the loads in-between, similarly to [83]. This hypothesis will be discussed in detail in section 4.2.

Just like dynamic programming methods, the event-based method does not ensure to find the global optimum and does not provide a measure of the solution quality. Approaches using the duality theory (Dual Lagrangian problem) formulate the dual problem, which is a bound to the optimal value of the primal problem, and thus a measure of the solution quality. The quality of the solution of the event-approach could thus be assessed by comparison to the equivalent mixed integer programming solution. This aspect will be discussed in the model validation section 2.2.

Stage

Initial stage

Stage k

Final stage

Algorithm

Define $\{E_0 = (e_1, ..., e_K)\}$; with $e_k \in \mathcal{E} \forall k \leq K$

For each stage $k \leq K$

Select the set of feasible states $\{I\} \subset \{J\}$

For each state I in $\{I\}$

Perform the local events optimization

Calculate the solution quality by integrating eq.(9)

last I in $\{I\}$?

no

$++I$ in $\{I\}$

yes

Define I_{opt} as the singleton $\{E_k\}$ which maximizes the profit at stage k

$k < K$?

no

$++k$

yes

$\{E_K\}$ local optimum solution

Method

Clean spread method. The events in E_0 define the K problem stages

Each modification in $\{E_{k-1}\}$ generates an element J in $\{J\}$. The state machine defines if $J \subset \{I\}$

The event list with best objective function is selected and defines the event list at next stage $\{E_k\}$

The event list I_{opt} at last stage is chosen as a local optimum to the problem

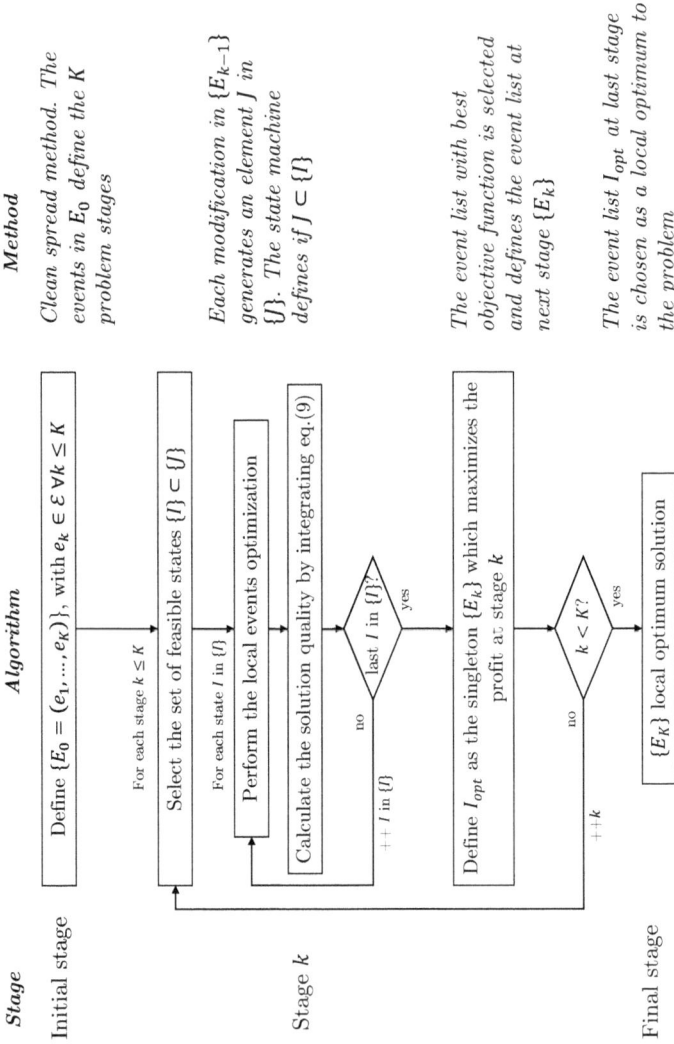

Figure 7 Flowchart describing the algorithm performing the event-based optimization of the self-scheduling problem

2.1.4 Event space optimization versus dynamic programming and timed automata

The vocabulary of dynamic programming has been purposely used, as both approaches have similarities. Dynamic programming decomposes optimization problems into sub-problems and uses a recursive approach to find the problem solutions using the solutions to the sub-problems [127]. Sequential decision problems are modeled using states and stages. States are defined so that the contribution to the objective function at a given period depends only on the previous state and the decision variables at that period [127]. Stages denote the sequences of the decision-making process [128]. In the literature, dynamic programming has been extensively used to solve the unit commitment and economic dispatch problem of an energy system (see the reviews [91, 93, 98, 129, 130]. To the best of the author's knowledge, the single unit self-scheduling problem from a price-taker perspective has not been solved using dynamic programming. The first noticeable difference between dynamic programming and the event-based approach lies in the time representation. The stages do not correspond to hourly time steps in the time horizon but the events in a dispatch. Where an optimization state at a given stage corresponds to the commitment status of all units at a given time step, the event optimization states correspond to the event-lists defining a possible operation profile. As dynamic programming suffers from the curse of dimensionality, the phase-space reduction heuristic is critical. The state machine semantic might be understood as a knowledge-based approach to reach such a phase-space reduction. The exclusion of all branches which do not lead to the best profit at each stage might be understood as a search window for truncation [130] phase space reduction strategy.

The vocabulary of the event-based approach is also similar to the timed automata theory, primarily via the definition of finite state machines. Automata theory is defined in [116] as "the study of abstract computing devices, or "machines""" and was initially used to model the human brain and define the tractability of problems for computing machines (Turing machine) [116]. A timed automaton is defined as a set of internal variables, states, start states, external and internal actions, discrete transitions and trajectories [131]. In the state machine defined for the event-optimization framework, the controlled deterministic transitions like start-ups and shutdowns correspond to trajectories in the timed-automata framework. The review in Part A section 3.3.3 has shown that automata for unit commitment and economic dispatch requires introducing the notion of state-dependent costs (priced timed automata) and are solved using model checking approaches, which have not been selected here.

2.2 Model validation and valuation

Just like for dynamic programming, the event-based approach does not provide a measure of the solution quality, as Lagrange relaxation does. To validate the approach, the results found by the event-based optimization are compared to the solution found by the approach presented in a scientific paper solving the same kind of problem.

2.2.1 Selected MIP formulation of the self-scheduling problem

The focus of this section is not on the MIP formulation so that the equations developed in the model of [103] are used and referred to as *TimeDiscretL*, see subsection 1.2.2. This formulation has been chosen due to its tightness, accurate differentiation between power and energy and possibility to integrate as many transitions as required. Moreover, the model accounts for the profit generated by the electricity during start-up.

2.2.2 Corresponding state machine for the event approach

Formulating the self-scheduling problem found in [103] into the event-based formalism requires the definition of a state machine. A possible state machine, applicable to conventional and combined cycle power plants without any combined heat and power, is shown in Figure 8. The only power plant variable accounted for in the selected MIP approach is the plant load [103]. Therefore, the vector x resumes to the plant load p, so that the state machine directly describes the load changes. The possible power plant states are (i) available, where the plant load remains constant at zero and (ii) operating, where the plant load remains constant. This constant load can be changed via the transition $r-> r$, for which the function limits the load change by the maximum allowed ramp-rate, which is the same for ramping up and down. The operating state also allows the transition to the available state via a desynchronization. This shutdown load curve is supposed to be known and independent from any further parameter. In [103] it is a linear decreasing function of time, so that this transition's load variation is of the form $\dot{x} = RR_{SD}$, with RR_{SD} the ramp-rate when shutting down. It is assumed that the synchronization transitions from the available state to the operating state depend on state values, namely the temperature range of the turbine rotor and boiler pressure parts. As no temperature variable is defined in the selected MIP approach, this temperature is transformed into a cooling duration range since last shutdown. Similarly to [103], five temperature ranges are defined with the corresponding start-up load curves. As for the shutdown, the start-up load curves are linear increasing functions of time, with the constant RR_{SU} being temperature dependent. These parameter values are defined in the next section.

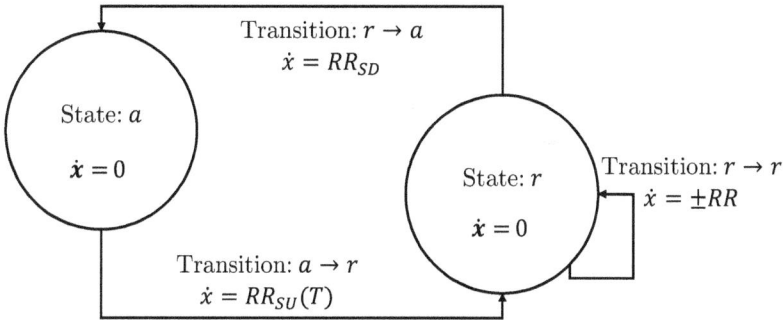

Figure 8 Power plant's state machine corresponding to the power plant description in [103]. The transitions are described by the load variation, which is limited by the respective ramp-rates. For start-ups, the ramp-rate limit is temperature dependent.

2.2.3 Parameter values

The parameters of the proposed test cases are assigned values similar to the test carried out in [103]. The start-up and shutdown economic and technical parameters are the same, and the curves have an hourly linear evolution of the load between $0\,\text{MW}$ and $\tilde{P}\,\text{MW}$. The remainder of the parameters are described in Table 9. The tested electricity price signal (taken from [103]) is to be found in Table 10. The parameter C^{LV} is the production cost and C^{NL} is the no-load cost. It must be noted, that the heuristic approach does not require the artificial definition of minimum up- and down-times TU and TD. The duration of an operation at full load or the offline duration result from the techno-economic trade-off between the costs of bringing the plant online and offline and the benefit obtained during the uptime. These events are only limited by the technical constraint that the power plant must be started until the minimum technical load at least (it cannot start until a load below it and then shut down again) and the logical constraint that a start-up cannot begin before the previous shutdown has been completed.

The uniform time discretization of the MIP problem is hourly. The nature of the event-based optimization does not allow the definition of a time step in the same sense, but to make the comparison as valid as possible, the heuristic is restricted to define events with t_k at full hours only.

The MIP tests have been carried out using Matlab R2014b and CPLEX 12.1. The MIP problems are solved to optimality. The event optimization algorithm has been programmed in C++. All tests have been carried out on an Intel® Core ™ i5 CPU M520 with 2.4 GHz and 4 GB of RAM.

Table 9 Value of the technological and economic model parameters for the event-approach validation

Technologic power plant description					
C^{LV} [€/MWh]	C^{NL} [€/h]	$\bar{P}$ [MW$_{el}$]	$\breve{P}$ [MW$_{el}$]	TU/TD[h]	$RR = RU =$ RD [MW$_{el}$/h]
55	200	378	150	3	76
RR_{SD} [MW$_{el}$/h]	RR_{SU1} [MW$_{el}$/h]	RR_{SU2} [MW$_{el}$/h]	RR_{SU3} [MW$_{el}$/h]	RR_{SU4} [MW$_{el}$/h]	RR_{SU5} [MW$_{el}$/h]
114	228	114	76	57	45.6

2.2.4 Model validation

Figure 10 illustrates the results of both the MILP and event-based approach on the described power plant model for the price signal 1. The hourly load levels resulting from the MILP optimization are plotted along with the load profile resulting from the sequence of events. The energy resulting from the chosen integration method is also represented. As visible in Figure 10, the event approach takes full advantage of the ramping ability of the plant since it ramps until 264 MW instead of remaining at 226 MW as imposed by the hourly time step of the reference model. This issue has also been identified in [132]. The event approach knows the end and the beginning of the ramp-down and ramp-up events respectively, as well as the ramp-rate allowed when operating above minimum technical load. The event approach is thus not tied to the time discretization and offers a different approach to the problem, here delivering a better dispatch than the reference model.

The test cases carried out in section 1.3 have been solved using the event-based optimization approach for the parameter values in Table 8. The problem formulation analyzed in section 1.3 includes non-linearity when the parameter cm_1 is strictly positive. This is the case in the Tests 4, 5, 6 and 9. The relative change in profit found via the event-based approach compared to the one found via the MIP approach is illustrated in Figure 9. The heuristic approach does not ensure to find the global optimum but the analysis shows that the gap is often negligible or even compensated by the better use of ramping capabilities between the hourly time steps.

Table 10 Tested electricity price signal taken and adapted from [103] in €/MWh

0	0	0	0	0	0	0	0	0	0	0	0	0	0	0	0	0	0	0	0
40	31	27	25	26	29	35	43	54	68	71	58	50	45	42	41	43	47	51	65
96	102	66	51	40	31	27	25	26	29	35	43	54	68	71	58	50	45	42	41
43	47	51	65	96	102	66	51	0	0	0	0	0	0	0	0	0	0	0	0
0	0	0	0	0	0	0	0												

Table 11 Comparison of the profits resulting from the MILP and event-optimization approach

Solving approach	Profit for hourly dispatch [€]	Relative profit change compared to the reference [%]
MILP	54,585	-
Event Optimization	55,194	+1.12

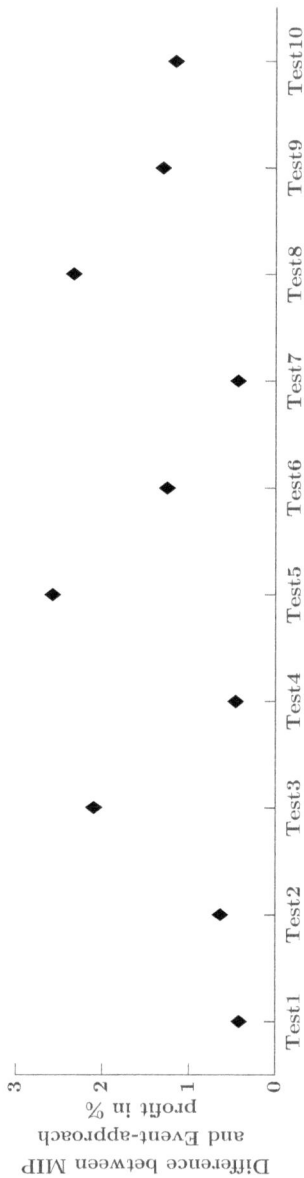

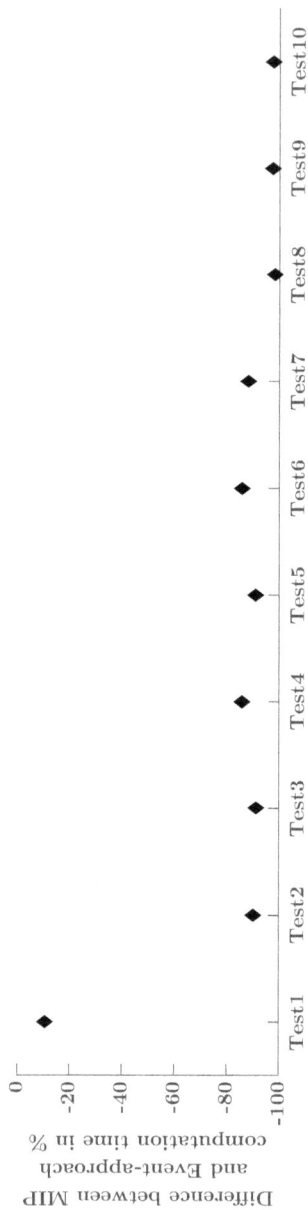

Figure 9 Change in profit (top) and computation time (bottom) found via the event approach compared to the one found via the MIP approach for different parameter values to be found in Table 8

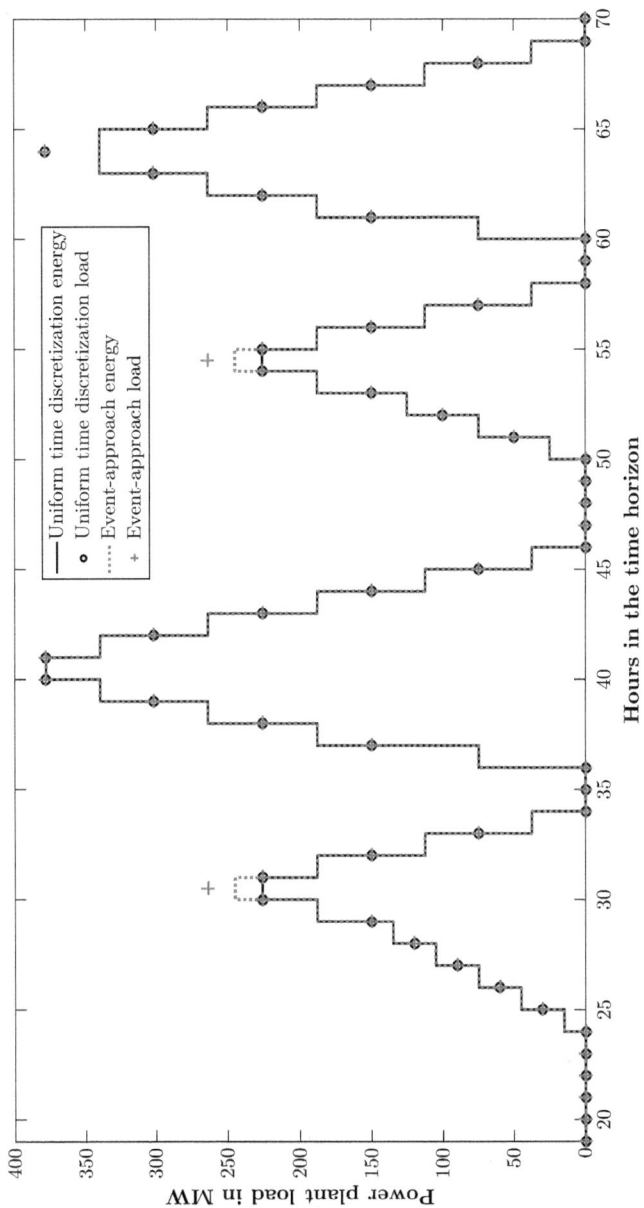

Figure 10 Comparison of the dispatch resulting from the MILP with the dispatch resulting from the event approach for the price signal 1

2.2.5 Computation time

Figure 9 displays the relative change in computation time observed during the various test case calculations. The *price signal 1* has been used to define a test case with a time horizon of a year, by iterating the price signal. In the average (twenty runs), the calculations are three times faster with the event-based approach algorithm than with the MILP approach solved to optimality. In the case of an optimality gap of 0.01 %, the calculation is two and half times faster on average. The event approach is preventing too large optimization spaces by using problem-specific model definitions. The MIP approach solution space is related to the whole amount of time steps in the considered time horizon, whereas the developed heuristic approach has a reduced but problem dependent solution space. The case of higher resolutions is to be further considered, such as the intraday market, for which the price changes every 15 minutes. Using the same price profile *price signal 1*, but with 15-minutes resolution instead of an hourly resolution, the MIP approach would need to address four times more points in the corresponding time horizon than with the hourly resolution. The heuristic approach would detect the same up and down events in its first step as with the hourly approach. The scaling is, however, related to the price signal. Considering the German day-ahead spot prices 2016, for a power plant with 20 €/MWh marginal operation costs, the first guess event list would have included a maximum of 178 start-up events. For marginal operation costs of 30 €/MWh (resp. 40 €/MWh), these would have been maximum 451 (resp. 219) start-up events. The phase space reduction can therefore not be systematically ascertained. It is interesting to note that a lot more start-up cycles have been detected at the power plant with 30 €/MWh marginal operation costs than the two other ones with 20 €/MWh and 40 €/MWh. The day-ahead prices in Germany in the year 2016 variate around an average value of 29 €/MWh. The power plant with 30 €/MWh marginal operation costs is thus more affected in terms of cycling by the oscillations around this mean value than the power plants operating at significantly lower or higher marginal operation costs.

2.3 Implementation

This section details a possible algorithmic implementation of the approach developed in the previous section to detect start-up, shutdown and part-load events defined in the state machine. An algorithmic modification of an event list (state) at each iteration (stage) is further proposed.

2.3.1 Detection of start-up and shutdown events

The first guess event list contains start-up and shutdown events. The detection of these events is based on the spark spread calculation. For an offline power plant (respectively in operating mode), when the marginal operation costs become greater (respectively lower) than the electricity price the plant should shut down (respectively startup). With the techno-economic power plant model introduced in this work, the condition is the change in sign in equation (10). The possible start-up and shutdown events are determined as follows: every time t_k the function becomes strictly positive a possible start-up event is detected and every time $t_{\overline{k}}$ the function becomes equal to 0 a shutdown event is detected.

$$(y(t) - cm_1) \cdot \overline{P} + y_0(t) - cm_h \qquad (10)$$

2.3.2 Detection of part-load events

The state machine further defines part-load cycling as possible events. Part-load operation is an economical option when electricity prices become infra-marginal for a duration and a price level that makes a shutdown and start-up more expensive than operating the plant at part-load. Like start-up decisions, part-load decisions are driven by the power plant model presented in this work. Figure 11 shows the evolution of the profit function, with arbitrary different electricity prices and thus y-values. When the condition 11 is true, profit can be made from a load defined in equation 12 on. However, it is clear that a larger profit can be made by operating the plant at full load, since the maximum efficiency is reached.

$$(y(t) - cm_1) \cdot \overline{P} + y_0(t) - cm_h > 0 \qquad (11)$$

$$p_m = -\frac{y_0 - cm_h}{y - cm_1} \qquad (12)$$

$$y - cm_1 < 0 \qquad (13)$$

When the condition 11 is false, two situations are to be distinguished: when the condition 13 is false, no profit is made, but the full load is still more profitable than part-load. When the condition (13) is valid, losses are lower at part-load than at full load. This way the criteria for part-load decisions writes as in (13).

2.3.3 Event list modification

The algorithm solving the scheduling problem requires a modification of the event list during the optimization process. These modifications are constrained by the state machine defined in Figure 8. The cooling and banking problem is here defined as determining whether (i) operating the plant at full load or (ii) not operating it at all or (iii) operating at part-load is more profitable. At a given stage, the three event modifications are thus to (i) not modify the dispatch, (ii) suppress a start-up and shutdown event sequence or (iii) replace a shutdown and startup event sequence by a ramp-down and ramp-up event sequence. The three states to compare are schematically pictured in Figure 12. The cooling and banking problem is solved at stage i by calculating the profit for the three options. The most economical option is chosen (depicted in black in Figure 12) and the next cooling and banking problem at stage $i + 1$ is solved.

2.4 Section conclusion

The proposed approach overcomes the uniform time discretization approach by the use of variable time discretization and state-machines inspired from the automata theory. Compared to the state-of-the-art MILP problem formulation, this concept makes better use of the power plant capabilities. A problem from the literature has been replicated for validating the approach, yet the power plant description is insufficient regarding operational flexibility. A better power plant description such as the one introduced in section 1induces larger problem sizes and eventually non-linearity. The calculations show that for this model, the event optimization approach does also provide better results and calculation times than state of the art solvers. These findings provide the basis for real-world operational flexibility assessments, which will be worked out in the case studies (Part C). In the next chapter, the problem formulation is enhanced by including the markets rewarding operational flexibility.

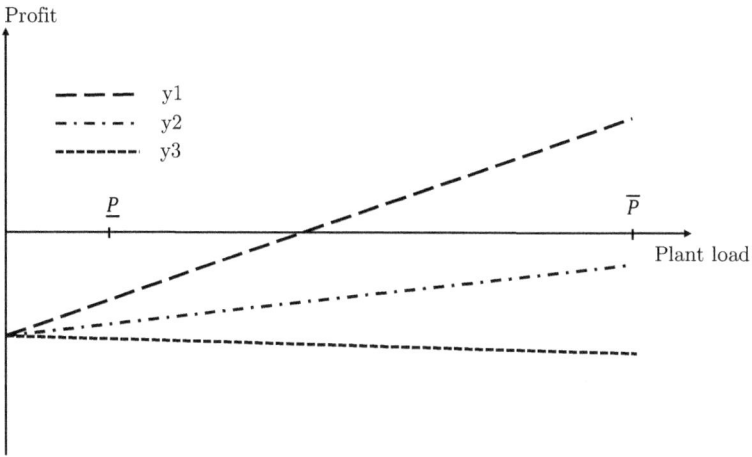

Figure 11 Evolution of the profit function for different electricity prices illustrating the part-load operation criteria

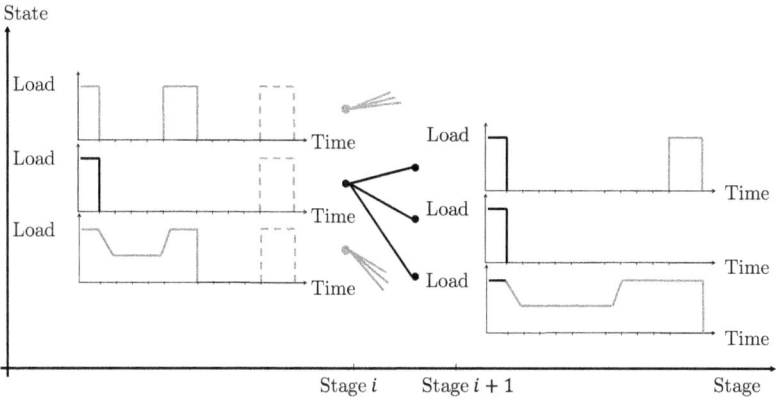

Figure 12 The cooling and banking optimization problem for the Event-list modification at each state

3 Assessing flexibility incentives in interdependent markets

In order to provide a suitable methodology for the assessment of the incentives sent by the liberalized energy markets, it is necessary to include the relevant markets in the model. The first section of this chapter provides an analysis of the markets thought to incent operational flexibility and a literature review on how these markets are included in unit commitment and economic dispatch models. Strategies for the implementation of these markets in the model developed in this work are then detailed. As the aim of this work is not to simulate the bidding behavior of the market actors, the price-taker assumption is still applicable.

3.1 Markets incentivizing operational flexibility

3.1.1 Markets and their design

The tasks to be performed by a power plant owner and operator are to plan and build power plants, to operate the power plant to deliver the contracted energy and to provide system services [79]. Energy can be contracted bilaterally over-the-counter or via power exchanges, generally divided into day-ahead and intraday but can also be part of power purchase agreements. System services are divided into frequency control, voltage control and power generation establishments like black start and island operation capability. In Europe, frequency control only is organized as a market.

In [27] from page 59 on, the authors characterize the institutional and market design related enablers for operational flexibility. Short time intervals and lead times (also identified in, e.g. [133]) are thought to unlock flexibility potentials. Short lead times are helpful as the forecasts gain in certainty as these get closer to real-time operation. Shorter time intervals allow counterbalancing the intra-hourly load variability. It is shown in [27] that larger balancing areas, shorter lead times and shorter time intervals lead to a reduced need for operating reserves. A further design question considers the existence of price caps and floors. Removing caps to allow price spikes in case of extreme events is necessary to incent flexibility. Undistorted price signals are a prerequisite if the market is to send the dispatching signal to the market participants. These price spikes are not necessarily impacting the end-consumer bill if appropriate hedging is put in place. [27] also mentions dedicated flexibility products like the Californian independent system operator (ISO) flexible ramp product. In the United States, this market product has been proposed and implemented by the Californian ISO and the Midcontinent ISO. These market-based flexible ramping products have been reviewed in [28], and extend the capacity approach provided by reserves with a ramp capability. This product is intended to improve the market operation via its inclusion in the unit commitment problem formulation and requires the calculation of flexible ramp requirements [28]. These flexible ramping products are not part of an organized market but rewarded via opportunity costs considerations [28]. In [134], the authors propose a similar product but in a power-based framework (as opposed to energy block based formulations). In [135] and [136] it is proposed to include ramping costs incurred by generating plants in complex bids to be accounted for by the independent system operator in the clearing process.

3.1.2 Literature review: markets, dispatch, and operational flexibility

In [132], the authors propose three models to find the optimal power trajectory minimizing the difference between the market operator's energy program and the hourly average values of the actual power trajectory. Infeasibilities are resolved using the intraday markets. In the proposed approach, the system operator acquires the secondary reserve after the day-ahead market settlement. This assumption does not hold for European markets, as in Germany, for instance, the capacity is reserved for a whole week. [137] proposes a weighted goal mixed integer programming model for the same joint dispatch and feasibility purpose. The authors of [132] provide a literature overview and distinguish between the papers working from a system perspective and those from a market participant perspective. As the latest is also of interest for this work, this review is limited to the same portion of literature. The paper [138] addresses a pool framework, in which the markets for energy, automatic generation control (AGC) and reserve are considered simultaneously. A tool for the power generator to optimally determine its involvement degree in the energy and capacity markets is developed. In this paper also, the problem of bidding is left out of consideration, and it is assumed that accurate price forecasts are available. The generator is also unable to influence the market prices. [139] improves the latter work by introducing start-up types as a function of the cooling duration. [140] formulates a self-scheduling problem including the reserve deployment timeframe and the real reserve availability of the plant, as it depends on its actual schedule. This formulation assumes exogenously defined reserve requirements. Whereas the previous references solve the problem of the optimal power allocation between energy and capacity markets, [141] and [142] propose theoretic approaches for defining the optimal bidding strategy. In [143, 144], the pricing of reserves in a North-American context with locational marginal pricing is based on Lagrange multipliers. [144] proposes a simultaneous energy and reserves market clearing model also based on Lagrange multipliers and compares it to sequential market clearing approaches. None of the studies reviewed here and in section 2.1.2.3, Part A include both the intraday and frequency control market.

3.2 Frequency control capacity reservation - an opportunity costs approach

A previous version of this chapter has been partially presented at the 1st International Conference on Large-Scale Grid Integration of Renewable Energy in India.

In Europe, the markets for frequency control are contracted by the TSO and divided into capacity reservation and energy provision. The capacity reservation ensures the TSO that, if needed, the reserved capacity is available to provide the required amount of energy. The capacity reservation is of interest to the flexibility topic as it limits the operating range of the power plant and thus might modify the value of operational flexibility, a supposition which will be analyzed in the case studies. The interested reader might refer to [145] and [133] for more details on the rules and functioning of the frequency control markets. The roadmap proposed in [133] is of interest for Europe but also elsewhere, especially regarding control procurement by intermittent resources. The latter is facilitated by more frequent tendering and shorter time periods [133]. Shorter lead times present a further advantage. Reference [133] suggests conditional bidding, which could help reduce the system must-run capacities. For cost efficiency, [133] proposes marginal pricing instead of pay-as-bid auctioning. To reduce the frequency control costs, a liquid intraday market is required, as the capacity contracted on this market can replace the tripped capacity. A sufficient intraday liquidity reduces the need for forecast error-related frequency control [133]. Please refer to Table 8 in [133] for further policy recommendations.

The scope of this work is limited to control power and thus excludes imbalance settlement aspects. As for the day-ahead dispatch optimization, a price-taker perspective with perfect price foresight is assumed. Again, the aim is not to optimize the bidding process, but to optimize the reservation of capacity for a given capacity reservation price and a day-ahead price signal. The capacity reservation is settled and known before the day-ahead market auction (on the contrary, as the review in subsection 3.1.2 and section 0, Part A has shown, in some markets the energy and reserves are co-optimized). Therefore, as in [146], if capacity is reserved, the supplier is supposed to bid its must-run capacity at a price of zero on the spot market. This ensures that the power plant is within the merit order and able to get its must-run capacity rewarded at the day-ahead market spot price. In the following section, the parametric frequency control market model developed for this study is detailed. This model is used to formulate the capacity reservation problem to be solved. As the study is aimed at operational flexibility, the state of the art opportunity cost approach is improved by accounting for the respective market timescales and degraded part-load efficiency.

3.2.1 Nomenclature

PUS	Program unit size (month, week, day...)
d_i	Program unit size of market i in hours
i	Index of market
t_b	Absolute time of the beginning of the program unit
$t \in \Omega$	Instant time t in the time range Ω
$c_p(t)$	Power plant marginal operation costs at time t in €/MWh
$EP(t)$	Electricity price at time t in €/MWh
Δ_{pos}	Reserved capacity in the positive direction during d_i in MW
Δ_{neg}	Reserved capacity in the negative direction during d_i in MW
Δ	Reserved capacity during d_i in MW
x_{min}	Minimum power output of power plant in $\mathrm{MW_{el}}$
$c_i(t)$	Opportunity cost for market i at the instant t in €/MWh
$c_i(PUS)$	Opportunity cost for market i during d_i in €/MW/PUS
$y(t)$	Load proportional component of the objective function, as function of time t in €/MWh
$y_0(t)$	Load independent component of the objective function, as function of time t in €/h
CP	Market capacity reservation price in €/MW/PUS

3.2.2 Parametric frequency control market

Table 12 Frequency control market parametric model

	Description	Possible values	Exemplary value
Program unit size	Duration for which the service needs to be made available	Year, month, week, day	Week
Program unit begin	Beginning of the program unit	Year: first of January; Month: first day of the month; Week: Monday; Day: first hour 00:00	Monday
Minimum bid size	Minimum capacity to be bid in the market	Typically between 1 and 5 MW	5MW
Full activation time	Duration within which the full committed capacity needs to be provided	Typically between 30 seconds and 15 minutes	5 minutes
Peak and off-peak product	Boolean indicating whether there are peak and off-peak products	Germany: peak is between 8:00 and 20:00 for weekdays. The rest is off-peak. France: off-peak is weekends.	Yes
Symmetrical product	The service can be either symmetric, which means that positive and negative control is required or not symmetrical, which means that a choice between positive and negative control can be made.	Yes/no	No

The scope of this work is limited to control power. It includes operating as well as contingency reserves, spinning reserves, positive as well as negative reserves, all activation times and ways of activation (cf. [133], chapter 2.4 for terminology). The various frequency control markets in Europe show similarities in design but some differences in implementation. The review and lessons learned in Europe regarding public procurement auction design (pricing, lead times, bidding periods among others) are used to model a single organizational framework, which can be parameterized to model different markets. The resulting parametric model is displayed in Table 12. Following simplifications are made for this model:

- Only one frequency market can be defined at a time and is described by the parameters in Table 12
- Even if the market settlement rule is pay-as-bid, uniform pricing will be assumed and calculated as the weighted average of the scored prices

- For each program unit time, only two products will be differentiated, namely a peak product and an off-peak product if the market is symmetric, and four in the other case: peak negative, off-peak negative, peak positive, off-peak positive.

3.2.3 Capacity procurement problem

As for the day-ahead dispatch optimization, a price-taker perspective with perfect price foresight is assumed. For bidding optimization refer to [146]. Since perfect foresight is assumed in the day-ahead market, the capacity procurement decision can be determined using the electricity day-ahead prices. The control market i is determined by its program time unit d_i. In Germany, it is a week for the primary and secondary frequency control and a day for the tertiary control. For each period, the perfect foresight helps determining the opportunity costs and by these means whether the operator will or not offer reserves in this period. Two cases are to be differentiated: either the plant's marginal operation costs are above the day-ahead electricity price or, they are below. Using the terminology of [147], the former are extra-marginal, whereas the latter are infra-marginal. One in all the opportunity costs equal the incurred losses for extra-marginal plants and the foregone profits for the others.

In case of positive capacity procurement Δ_{pos} in the frequency control market, the plant has to operate at its minimum technical load x_{min} at least. For negative procurement the load level should be at least $x_{min} + \Delta_{neg}$. In the literature these opportunity costs do not take into account the increased operation costs at part-load; mostly constant marginal costs are assumed. With this assumption, the opportunity costs for extra-marginal plants are the losses allocated to the offered control capacity as in equations (14) for positive control and (15) for negative control. If the effect of the efficiency loss at part-load needs to be taken into account, as well as the maintenance costs, the power plant model introduced in the section 1.1 can be used. The losses can then be written as in equation (16) for both control directions and for a given load level p. The power plant's profit function here includes the profit and costs related to the combined provision of heat, see section 3.4 for more detail. In order to make the equations more compact, the new functions (17) and (18) are introduced.

$$c_i(t) = (c_p(t) - EP(t)) * \frac{x_{min}}{\Delta_{pos}} \qquad \forall t \,/\, EP(t) < c_p(t) \qquad (14)$$

$$c_i(t) = (c_p(t) - EP(t)) * \frac{x_{min} + \Delta_{neg}}{\Delta_{neg}} \qquad \forall t \,/\, EP(t) < c_p(t) \qquad (15)$$

$$c_i(t) = -\frac{\tilde{y}(t) * p(t) + \tilde{y}_0(t) + y_h(t) * h(t)}{\Delta} \qquad \forall t \,/\, EP(t) < c_p(t) \qquad (16)$$

$$\tilde{y}(t) = y(t) - cm_1 * u(t) \qquad \forall t \qquad (17)$$

$$\widetilde{y_0}(t) = y_0(t) - cm_h * u(t) \qquad \forall t \qquad (18)$$

In case the marginal operation costs are below the day-ahead electricity price, the opportunity costs are the foregone profit allocated to the control capacity. The positive control capacity reserved in the control market is not available for the day-ahead market anymore. This way the capacity costs are equal to zero when the day-ahead electricity price equals the marginal costs of the plant, see equation (19). In case of negative control power, the infra-marginal power plant has no opportunity cost, see equation (20). If the effect of the efficiency loss at part-load needs to be taken into account, the foregone profits for positive capacity can be written as in equation (21).

$$c_i(t) = \frac{(EP(t) - c_p(t)) * \Delta_{pos}}{\Delta_{pos}} \qquad \forall t \,/\, EP(t) \geq c_p(t) \qquad (19)$$

$$c_i(t) = 0 \qquad \forall t \,/\, EP(t) \geq c_p(t) \qquad (20)$$

$$c_i(t) = \frac{\tilde{y}(t) * \Delta_{pos}}{\Delta_{pos}} \qquad \forall t \,/\, EP(t) \geq c_p(t) \qquad (21)$$

The decision to participate or not in the market during the period d_i is defined by the comparison of the opportunity costs (calculated via the day-ahead electricity price and operation costs) and the capacity prices on the frequency market (perfect foresight is also assumed). This way the participation decision criteria is submitted to the comparison of the opportunity costs c_i and the profits on the control market, which is the capacity price CP. The opportunity costs are calculated for a given day-ahead electricity price, which means that these costs are valid for a given hour only. The decision criteria for the whole program unit size should then write as in equation (22).

$$c_i(PUS) = \frac{1}{PUS} \int_0^{PUS} c_i(t) \, dt < CP \qquad (22)$$

3.2.4 Quantitative analysis of the improved procurement problem

To quantify the effect of the improved problem formulation, an exemplary opportunity cost calculation is performed. The calculations are made using the data of the first week of the year 2014, from January 6th to January 12th. The capacity reservation price signal used in the following is the weighted average of the secondary frequency control market results in Germany, as the market uses pay-as-bid pricing. There are four products to be accounted for, as the market is symmetric (negative and positive control) and as there are two tariff periods (high and low tariff). The electricity price signal is the German day-ahead spot market price. The power plant is assumed to have 26 €/MWh marginal operation costs, a minimum complaint load at 150 MW for a nominal load of 378 MW and a ramp rate of 3 %/min. An 8.8 percentage point efficiency reduction at minimum load is assumed. Figure 13 compares opportunity costs without taking into account the efficiency losses at part-load and any other dynamic effect like cycling costs (referred to as static) and the same opportunity costs with these parameters (referred to as dynamic). Due to the higher activation time in the tertiary control market (mFRR), more capacity can be offered, and thus the opportunity costs are lower than the secondary control market opportunity costs. The higher opportunity costs in the out-of-merit case show that these are more negatively impacted by the spinning capacity reservation than by the foregone profits in the day-ahead market. The results also demonstrate that neglecting operational flexibility parameters means underestimating the opportunity costs of frequency control capacity reservation. The improved positive control opportunity cost calculation does not lead to a zero opportunity cost when the day-ahead price equals the marginal operation costs, as the spinning control still leads to losses due to the degraded efficiency. As soon as the day-ahead prices are strictly higher than the marginal operation costs, the opportunity costs equal the difference in operating costs due to the efficiency loss at the operated load, which is the nominal load minus the reserved capacity.

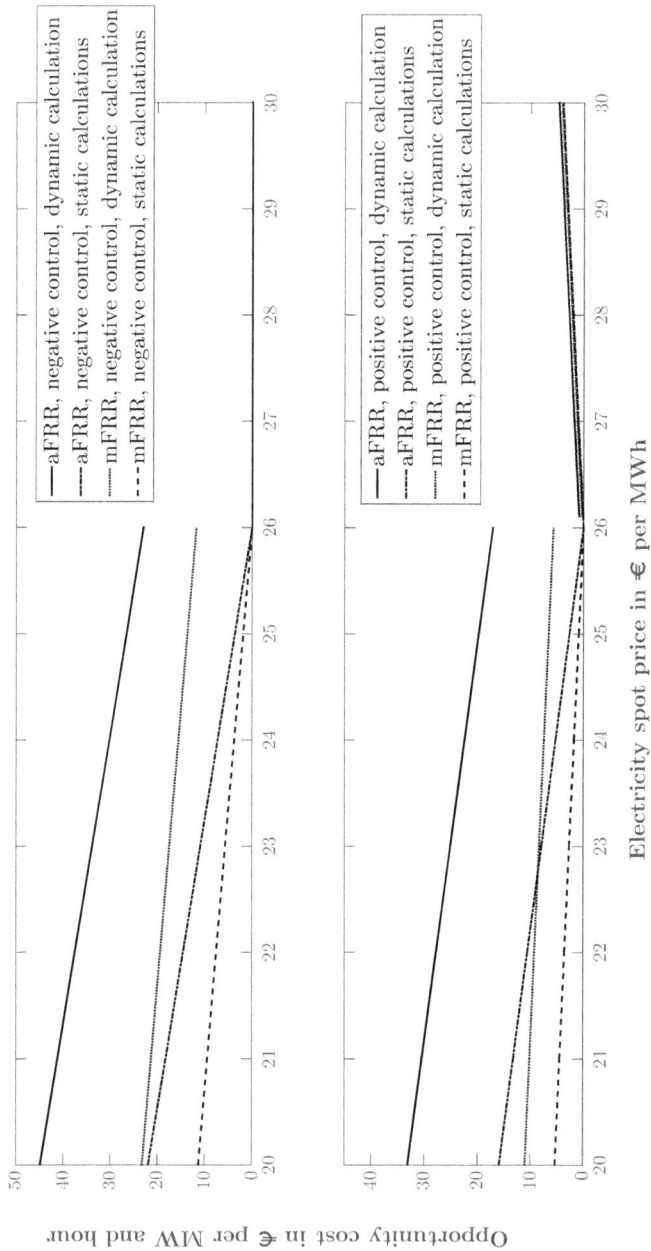

Figure 13 Comparison of opportunity costs for negative (top) and positive (bottom) frequency control capacity provision when accounting for dynamic effects or not. aFRR: automatic frequency restoration reserve. mFRR: manual frequency restoration reserve.

3.2.5 Energy procurement problem

Need for control power stems from system imbalances, which are caused by stochastic events such as load forecast errors and unplanned outages and deterministic events like schedule leaps. For the energy procurement problem, a stochastic event should be introduced with a given call probability function. Adding this stochastic energy provision to the model would provide different results at each call of the same job. Moreover, the rewards for providing energy are much more significant than those for capacity provision so that the profit is likely to be improved through the energy provision. For the assessment, it is therefore preferred not to add the energy provision to the dispatch.

3.2.6 Implementation

Provision of frequency control services is submitted to strict requirements, mostly regarding the ability to provide specific ramprates in a given amount of time. Therefore, it is first necessary to verify whether the generating plant is flexible enough to contribute to this market. This is achieved by calculating the maximum capacity that can be offered (ramp-up and ramp-down rate times the full activation time) and comparing it to the minimum bid size of the market.

The capacity procurement auction takes place before the day-ahead spot market. The proposed approach does not optimize the bidding process in the frequency control market. This means that it does not optimize in which markets (primary, secondary, and tertiary) the operator should bid, and it does not optimize which amounts and prices should be offered. See [146] and [148] for bidding strategies in the frequency control market. The reserved share is bid at a price of zero in the spot market. This will have the impact to reduce the outcome of the spot day-ahead market in times of low demand, and to increase it in times of high demand, as demonstrated in [146]. However, these interactions are not taken into account in the presented model, due to the infinite market assumption. All market signals are provided exogenously. The developed approach calculates for each program unit time whether the operator should or not reserve capacity and optimizes the plant operation accordingly. For a given market, if the products are non-symmetric and if there are peak and off-peak products, the algorithm, additionally, optimizes for which products capacity should be offered.

Spinning control introduces a must-run condition, which modifies the topology of the state machine. This must-run condition is formulated using the conditions (23), (24) and (25). The start-up and shutdown decisions are only restricted at times when frequency control capacity is reserved so that the conditions only impacts times between $[t_b, t_b + d_i]$, with t_b the beginning of a new program unit d_i. Start-up decisions at time t are submitted to condition (23) and (24). The validity of condition (23) defines that a start-up is required if control capacity is reserved, even if the day-ahead prices are not high enough to make profit. The validity of condition (24) allows the plant to start-up if the electricity price is high enough for profits when no control capacity is required.

$$\Delta(t-1) = 0 \quad \& \quad \Delta(t) > 0 \quad \forall \tilde{y} * \overline{P} + \widetilde{y_0} \tag{23}$$

$$\tilde{y} * \overline{P} + \widetilde{y_0}\,(t) \geq 0 \quad \& \quad \tilde{y} * \overline{P} + \widetilde{y_0}(t-1) < 0 \quad \& \quad \Delta(t) = 0 \tag{24}$$

Shutdowns are not allowed anymore when spinning frequency control capacity is reserved, i.e., when $\Delta(t) > 0$. Shutdown decisions at time t are submitted to the validity of condition (25), which means that the power plant can only shutdown when the day-ahead prices goes below the profit limit and no control capacity is required.

$$\Delta(t) = 0 \quad \& \quad \tilde{y} * \overline{P} + \widetilde{y_0}\,(t) < 0 \quad \& \quad \tilde{y} * \overline{P} + \widetilde{y_0}(t-1) \geq 0 \tag{25}$$

3.2.7 Conclusion

The presented concepts and implementation allow optimizing the participation to capacity reservation for frequency control in European markets from a price-taker perspective. The European markets for frequency control procurement have been parametrized to a single model. The method is based on an opportunity cost approach and improves the state of the art equations to include part-load efficiency degradation and the different time scales of the interdependent markets.

3.3 Quantitative assessment of arbitrage opportunities in the intraday market

In the intraday market, power plants can correct the position contracted on the day ahead market. The power plant flexibility limits the participation to the market in the magnitude of the ramp-rate capability. Therefore, the impact of the power plant flexibility on the revenues in the intraday market (and frequency reserve market) needs to be analyzed. The intraday scheduling requires continuous re-optimization since information actualizes and becomes more accurate the closer the gate closure. As in the case of the day-ahead dispatch, options are determined regarding the state of the plant and the price spreads, thus defining the power plant's arbitrage possibilities in the intraday market. High intraday prices are a signal of a system requiring increased power production. This can result from lower renewables infeed than predicted or unplanned unavailabilities. In this case, plants are encouraged to increase their power output in the intraday market. In case of low intraday prices, the signal indicates an oversupply caused by a renewables infeed higher than planned. In this case, operators are encouraged to reduce their power production since they can realize their day-ahead position with cheap intraday power.

3.3.1 Qualitative approach

Table 13 qualitatively summarizes the different options in the intraday market depending on the day-ahead schedule and the intraday price forecast. For instance, if the power plant is operating and scheduled to remain online (on→on) and the intraday price forecast is high, then the load should be increased to sell more electricity on the profitable intraday market. This increase is limited by the nominal load capacity of the plant (eventually minus the positive control power positions) and the power plant ramp-rate. In the same case but with low intraday prices, the power plant is encouraged to reduce its production. Low intraday prices are the opportunity to buy this cheap energy to provide the energy sold on the day-ahead market. This is limited by the minimum load plus eventually negative control power positions and the power plant ramp-rates. In case the power plant is offline and scheduled to remain so (off→off), high intraday prices are incentives to operate the plant to sell the missing energy (not provided by the renewables). In case the plant is scheduled to start to provide energy on the day-ahead market (off→on), but the intraday forecasts are low, the strategy of buying this cheap energy could be applied. On the contrary, high intraday price forecasts are incentives to start the power plant sooner, to sell the energy on this market. In case shutting down the power plant is planned (on→off), high intraday price opportunities can be caught by postponing the transition. In case of high intraday prices and a power plant in its "available" state, the arbitrage opportunity can only be taken if the price signal is forecasted long enough in advance for the power plant to start (typically 4 to 12 hours for a thermal power plant).

Table 13 Options for intraday modifications of the day-ahead dispatch

	Scheduled start off→on	Scheduled shutdown on→off	Operating on→on	Available off→off
Low intraday price forecast	Cancel or postpone start	No change	Decrease load	No change
High intraday price forecast	Start sooner	Cancel or postpone shutdown	Increase load	Start-up and shutdown

3.3.2 Quantitative approach

$p_{DA}(t)$ Power plant load in the day-ahead dispatch in MW

$p_{ID}(t)$ Power plant load in the intraday dispatch in MW

$p(t)$ Power plant load in the combined day-ahead and intraday dispatch in MW

$P(t)$ Profit in the combined day-ahead and intraday dispatch in €

$P_{DA}(t)$ Profit in the day-ahead dispatch in €

$P_{ID}(t)$ Profit in the intraday dispatch in €

$EP_{DA}(t)$ Electricity price in the day-ahead market

$EP_{ID}(t)$ Electricity price in the intraday market

$\tilde{y}_{ID}(t)$ Value of the $\tilde{y}$ function in the intraday market, load independent component of the profit function, as function of time t in €/h

$\tilde{y}_0(t)$ Value of the $\widetilde{y_0}$ function in both the intraday market and day-ahead market (as it is price independent), Load independent component of the objective function, as function of time t in €/h

Figure 14 compares the self-scheduling of a given power plant in the day-ahead (DA) market and the intraday (ID) market. Both dispatches are optimized independently and are used for the illustration of the following concept. The boxes in Figure 14 illustrate three different behaviors, which will be used to modify the day-ahead dispatch to catch intraday arbitrage opportunities. If the day-ahead and intraday dispatch are the same, as in the first box in Figure 14, then this dispatch is operated, and the profit is the one resulting from the day-ahead prices. This case is the one where profit can be made on the intraday and the day-ahead market, and where a strategic capacity reservation for the intraday market could have resulted in an increased profit. If the day-ahead dispatch is offline while the intraday one is online, second box in Figure 14, then the intraday dispatch is operated, and the profit is the one resulting from the intraday prices. There is no day-ahead position to supply so that the whole energy is sold at intraday prices. If the day-ahead dispatch is online, but the intraday dispatch is offline or at lower load, then the intraday dispatch is operated. In this configuration, the day-ahead position is supplied using energy bought on the (at the moment competitive) intraday market. This way the operation costs result from the intraday dispatch and the energy bought at the intraday price and the profit results from the day-ahead position. The algorithm thus:

- First optimizes the day-ahead dispatch
- Then optimizes the intraday dispatch (independent from the day-ahead results)
- Then reschedules using the set of rules 26.

$$if\ p_{DA} = p_{ID}\ then\ p = p_{DA} = p_{ID}\ and\ P = P_{DA}$$

$$if\ p_{DA} < p_{ID}\ then\ p = p_{ID}\ and\ P = P_{ID}$$

$$if\ p_{DA} > p_{ID}\ then\ p = p_{ID}\ and\ P = (EP_{DA} - EP_{ID}) * p_{DA} + \tilde{y}_{ID} * p_{ID} + \overline{\tilde{y}_0}$$

$$(26)$$

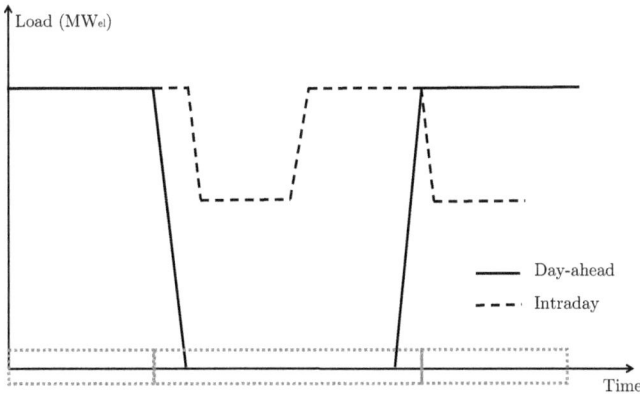

Figure 14 Day-ahead and Intraday dispatch for the combined dispatch in both markets

3.3.3 Intraday market price signal

In Germany, two intraday markets with different designs are implemented. The first one is a pay-as-bid continuous intraday market with hourly and 15-minute products and the second one is a fixed-gate auction with 15-minutes products with marginal pricing. The continuous pay-as-bid design means that there are as many prices as accepted bids, which requires defining a price for the model. Until 2015, the intraday index at the EPEX SPOT was defined as the weighted average of all realized prices. In July 2015, a new index, named ID3-Price was proposed, as the reference index for the German intraday cap future product. It is the weighted average of the continuous market prices realized the three last hours before delivery [149].

3.3.4 Conclusion

The operation in the intraday market has been modeled as an arbitrage opportunity between the day-ahead and intraday markets. This approach is in line with the contribution which is desired from this market from a system perspective: the correction of forecast errors and thus the reduction of real-time balancing needs. As in every market without a single price signal (continuous bidding, pay-as-bid), the price-taker perspective raises the difficulty of a suited price signal definition.

3.4 Combined heat and power plants

Some power plants deliver not only electricity but also heat to district heating networks or provide process steam for the industry. The review of flexibility providers in Part A section 1.2.2 has shown that combined heat and power (CHP) plants contribute to the energy transition by coupling the heat and electricity sector. At CHP plants, the production of heat and electricity is coupled unless a heat storage system is implemented. CHP plants exist in two variants; the back pressure turbine plant (non-condensing plants) and extraction-condensing turbine plants. The first ones have a strict power-heat relationship. Due to their higher flexibility, the extraction-condensing turbine plants are considered here and have the feasible operation regime depicted in Figure 15. The modification of the self-scheduling objective function due to the provision of heat will be detailed. This objective function might be used in a mixed integer programming formulation as well as in an event-based approach. It is assumed that the heat provision prevails over the electricity provision, so that combined heat and power introduces a must-run condition. The impact of this must-run condition on the state machine will be detailed in terms of conditions formulations, as it has been made for frequency control in subsection 3.2.6.

3.4.1 Nomenclature

3.4.1.1 Exogenous Parameters

$HP(t)$ Thermal heat price in €/MWh$_{\text{th}}$
$EP(t)$ Electricity price at time t in €/MWh
$FP(t)$ Fuel price including CO_2 emission costs at time t in €/MWh$_{\text{th}}$
s Loss of electricity generation per unit of heat generated at fixed fuel input in MW$_{\text{el}}$/MW$_{\text{th}}$; assumed constant.
c_{bp} Back-pressure coefficient in MW$_{\text{el}}$/MW$_{\text{th}}$
$h(t)$ Heat production in MW$_{\text{th}}$

3.4.1.2 Endogenous Parameters

$p_h(t)$ Effective electric power in MW
$p(t)$ Electric power in MW corresponding to the same operation point as $p_h(t)$ but without heat extraction
$f(t)$ Fuel consumption rate of power plant at time t, in MW$_{\text{th}}$
$c_m(t)$ Maintenance costs as function of time t in €/h

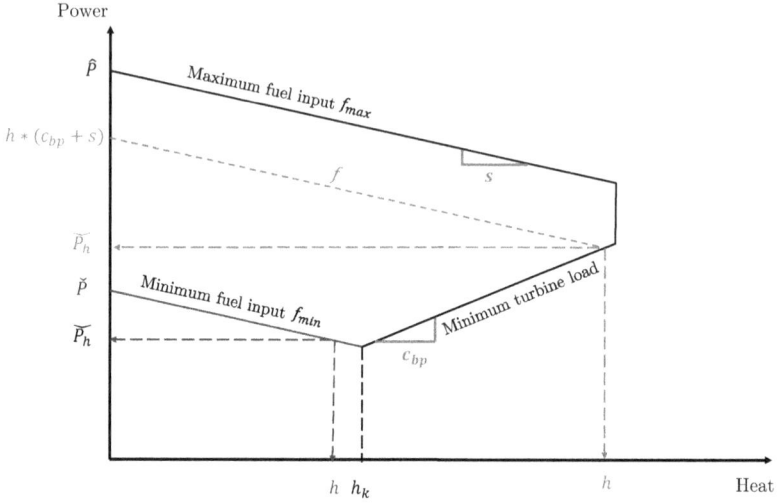

Figure 15 Combined heat and power plant feasible operation region

3.4.2 Implementation: Objective function

With the provision of heat, the self-scheduling profit maximization problem writes:

$$max_{p_h} \int_{\Omega} [\, EP(t) \cdot p_h(t) + HP(t) * h(t) - FP(t) \cdot f(t) - c_m(t)\,]\, dt \qquad (27)$$

Due to the combined production of heat and power, the load calculates as the difference between the load without heat provision $p(t)$ and the loss of load due to heat provision $* h(t)$:

$$p_h(t) = p(t) - s * h(t) = \bar{P} * \frac{f(t) - f_0}{f_1 - f_0} - s * h(t) \qquad (28)$$

The objective function becomes:

$$max_p \int_{\Omega} [\, EP(t) \cdot p(t) + (HP(t) - EP(t) \cdot s) * h(t) - FP(t) \cdot f(t)$$
$$- c_m(t)\,]\, dt \qquad (29)$$

Using the power plant model introduced in section 1.1, this becomes:

$$max_p \int [(\, y(t) - cm_1 * u(t)) * p(t) + y_0 + y_h(t) * h(t) - cm_h * u(t)$$

$$- \sum_{k \in L} f_k(t) \cdot \delta(t - t_k) \cdot FP(t) - \sum_{i \in E} ce_i \delta(t - t_i)] \, dt \qquad (30)$$

$$y(t) = EP(t) - \frac{(f_1 - f_0)}{\bar{P}} \cdot FP(t) \qquad (31)$$

$$y_0(t) = -(f_0 \cdot FP(t) + cm_0) \qquad (32)$$

$$y_h(t) = HP(t) - EP(t) \cdot s \qquad (33)$$

It is also worth noting that the lower bound of the feasible operation region (see Figure 15) imposes a distinction between heat extractions lower and greater than the limit h_k defined with the equation (34). Equations (35) and (36) show that the load that can be delivered at part-load when extracting the heat h depends on whether the heat extraction amount is less or more than h_k .

$$\check{P} - s * h_k = c_{bp} * h_k \qquad (34)$$

$$\widetilde{P_h} = \check{P} - s * h + \Delta_{neg} \quad \forall h < h_k \ and \ f \geq f_{min} \qquad (35)$$

$$\widetilde{P_h} = c_{bp} * h + \Delta_{neg} \quad \forall h \geq h_k \ and \ f \geq f_{min} \qquad (36)$$

3.4.3 Implementation: Must-run condition

The must-run condition introduced by the heat provision changes the topology of the state machine. This modification is handled via the conditions (37), (38) and (39). The conditions (37) and (38) define the start-up behavior. If the condition (37) is true, a start-up is required for the heat or frequency control provision, even if the day-ahead prices are not high enough to make profit. If condition (38) is valid, there is no heat or frequency control contract, so that the plant is allowed to start-up.

$$\Delta(t-1) = 0 \ \& \ \Delta(t) > 0 \quad or \quad h(t-1) = 0 \ \& \ h(t)$$
$$> 0 \quad \forall \tilde{y} * \overline{P} + \widetilde{y_0} \tag{37}$$

$$\tilde{y} * \overline{P} + \widetilde{y_0}(t) \geq 0 \ \& \ \tilde{y} * \overline{P} + \widetilde{y_0}(t-1) < 0 \ \& \ \Delta(t) = 0$$
$$\& \ h(t) = 0 \tag{38}$$

Shutdown decisions at time t are submitted to condition (39), which means that the power plant can only shutdown when no heat or frequency control capacity is required. The initial condition is determined by initial heat demand. Either the initial heat demand is positive so that the plant is operating or there is no heat demand and the electricity price determines the initial state.

$$\Delta(t) = 0 \ \& \ h(t) = 0 \ \& \ \tilde{y} * \overline{P} + \widetilde{y_0}(t) < 0 \ \& \ \tilde{y} * \overline{P} + \widetilde{y_0}(t-1) \geq 0 \tag{39}$$

3.5 Market-dependent value of flexibility improvements

The day-ahead, intraday and frequency control markets have been identified as the ones intended to incent operational flexibility. Given the different time scales, lead times and functioning of these markets, the hypothesis that the value of a given flexibility improvement depends on which markets and market combinations the power plant is dispatched in might be formulated. This section provides a quantitative test case.

3.5.1 Test case

The work at hand does not only allow for a quantification of the value of flexibility, but also for a quantitative assessment of operational flexibility improvements' value. The idea is to assess the value of upgrades, which improve the operational flexibility of a conventional power plant, and to base this assessment on the full operation profile. The operational flexibility upgrade value results from the comparison of the optimized dispatch of the baseline power plant with the optimized dispatch of the improved power plant. The comparison of the resulting dispatches would deliver insights into the value of the improvement in various aspects including the cycling behavior (load following and start-ups) and profit. Figure 16 illustrates the proposed method.

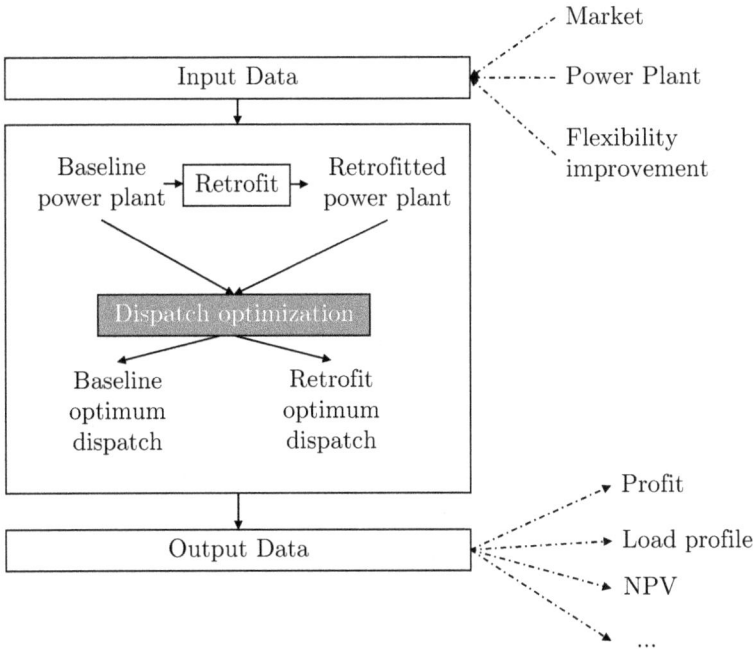

Figure 16 Unit commitment and economic dispatch for the valuation of flexibility improvements: illustration of the proposed concept

To demonstrate how the benefits of a given operational flexibility improvement are market dependent, a ramp-rate acceleration with reduction of minimum load and reduction of start-up fuel is assessed in the Spanish market for the years 2014 and 2016 at a supercritical hard coal-fired power plant. The simulation parameters are the same as in the case studies presented in the next part (power plant reference SC-H).

The power plant and its retrofitted version are dispatched in the Spanish market 2014 and 2016. The dispatch is optimized in the day-ahead market only and then in the day-ahead market with intraday arbitrage. A third calculation dispatches the plants in the day-ahead market with frequency control capacity reservation. The last calculation optimizes the dispatch in the three markets together. For each of these dispatches, the yearly profit is calculated. The increase in yearly profit due to the flexibility improvement is used as a metric.

3.5.2 Results

The results in Table 14 show that the intraday markets and the market for frequency control influence the value of operational flexibility. For this particular power plant and market environment, the results show that the operational flexibility product has more value in the year 2016 than in the year 2014 if no capacity is reserved for frequency control. The capacity reservation for frequency control, however, increases the value of the product, when compared to day-ahead only and day-ahead with intraday arbitrage. Intraday arbitrage also influences the value of the product when compared to the day-ahead market only, but in a smaller range than frequency control. In the year 2014, intraday arbitrage increased the day-ahead value of the product, whereas in 2016, it decreased it. The statistical analysis of the Intraday prices in the year 2014 and 2016 shows that the prices have less variability and lower average value in 2016 than 2014, which explains this result and might be due to the slightly reduced renewables infeed. This assessment demonstrates that comprehensive flexibility studies should include the intraday and frequency control markets. The market design itself further influences the results, as will be shown in the case studies.

Table 14 Value of a flexibility improvement in different flexibility markets in Spain 2014 and 2016 measured as the yearly profit increase in %

	Day-ahead	Intraday arbitrage	Frequency control capacity reservation	Intraday arbitrage and frequency control capacity reservation
Spain 2014	2.5%	2.9%	28%	29%
Spain 2016	5.6%	4.5%	11%	27%

3.6 Section conclusion

The markets intended to incent operational flexibility have been identified. The dispatch optimization algorithm has been improved to account for these markets. As the scarce literature shows, taking these markets into account is not a straightforward approach. In this work, the intraday market has been included as an arbitrage option and the frequency control capacity reservation as an opportunity cost optimization task. The method improves the state of the art equations to include part-load efficiency degradation and the different time scales of the interdependent markets. The European markets for frequency control procurement have been parametrized to obtain a single model. As in every market without a single price signal (continuous bidding, pay-as-bid), the price-taker perspective raises the difficulty of the price signal definition. The model has also been improved via the inclusion of the combined heat and power provision. A case study in the Spanish market environment has shown that the value regarding yearly profit increase is sensitive to which market the power plant is dispatched in, and thus that a comprehensive operational flexibility analysis should include these markets.

4 Applicability of the study assumptions

4.1 Merchant plant operating in forward and spot markets

4.2 Power plant technical description
4.2.1 Power plant ramp-rates
4.2.2 Minimum up- and downtimes

4.3 Intraday lead time

4.4 Fleet effects

4.4 Stochasticity

4.6 Section conclusion

This chapter raises the question of the validity and applicability of the most critical assumptions made for this work. The approximations and assumptions might lead to discrepancies between the power plant's real dispatch and the one obtained by solving the proposed problem. When comparing two operational profiles for flexibility valuation, these discrepancies stem from the same assumptions and thus do not impact the relative value of the operational flexibility increase.

4.1 Merchant plant operating in forward and spot markets

As noted by [50], market liquidity is required for market participation. Market liquidity is thus a validity indicator for the assumption that the power plant operators sell their production on the wholesale markets. The Office of Gas and Electricity Markets (OFGEM) defines the liquidity of a market as "the ability to quickly buy or sell a commodity without a significant change in its price and without incurring significant transaction costs" [150]. According to this definition, the liquidity or market depth (impact of volume on price) is also a measure of the validity of the infinite market assumption, which has been used to state that the dispatch decision of a single unit does not affect market prices. Another aspect highlighted by the measure of the market liquidity is how often and by how many players the market is used to trade goods.

The traded volume, the churn factor or diversity of parties are possible measures of the market liquidity. According to the European Union Electricity Market Glossary [47], in the context of energy markets, the churn factor is "a frequently used indicator of liquidity [...] (i.e., the number of times electricity generated in a market is subsequently traded). The churn rate is also calculated as the ratio between the volume of all trades in all timeframes executed in a given market and its total demand" [151]. "Another way of defining a churn rate is the volumes traded through exchanges and brokers expressed as a multiple of physical consumption" [47]. ACER and CEER [151] report the chunk factor and trading volumes on forward markets and ratio between traded volume and demand in intraday markets in Europe. The forward market's liquidity reported by [151] shows that Germany, Austria, Luxemburg, the United Kingdom, Nordic countries and France (since 2015) have satisfactory liquidity. Spain, Italy and the Netherlands do not. These conclusions are corroborated by the reported trading volumes [151]. In 2016 Spain, Italy, Portugal and Great Britain, followed by Germany/Austria/Luxembourg had the highest intraday liquidity [151].

With low demand elasticity and few competitors, the electricity prices are more likely to deviate from their competitive level [50], so that the number of competitors in the market is decisive. This aspect drives the validity of the assumption that electricity market prices are a suited input signal to the power plant operator's dispatch decisions. To further answer the question by how many players the market is used to trade goods, the European Commission uses indexes based on the Herfindahl Hirschman Index (HHI), an index based on the market share of market participants and which increasing value indicates increasing market concentration. Most of the European countries have increased competition in power generation, facilitated by the increase in renewables [152]. France, however, still shows high market concentration with one company owning 77 % of the generation capacity in 2015 [152]. Countries with an index lower than 2000 (moderate market concentration) are Germany, Denmark, Spain, Italy, Netherlands, Austria, Poland, Romania, Finland and the United Kingdom.

This index is also used in combination with other indexes to define the degree of liberalization of markets. In a liberalized market, power plant operators have to compete to maximize their profit. Fully merchant power plant assumptions might not meet some countries' current reality. Power purchase agreements with arbitrage of the remaining capacity on the spot markets might be closer to reality [153], and is referred to as semi-compulsory markets in the literature [48]. As argued in [154], the power exchange prices are setting a reference for contracts settled outside of the power exchanges (e.g., over-the-counter or futures), as no buyer would agree on these contracts if he could do better on power exchanges. As power purchase agreement data is not disclosed, to come closer to an answer to the question whether liberalized energy markets send the right incentives for flexible operation, fully merchant plant using spot market is a valid assumption.

4.2 Power plant technical description

4.2.1 Power plant ramp-rates

One of the event-based optimization phase-space reductions lies in the state machine itself, as it can be defined to allow transitions between minimum load and nominal load without considering the loads in-between, similarly to [83]. This section discusses why this is a valid approach. The required knowledge has already been worked out in section 2.3.2 in the form of a rule for incentives to operate at part-load. Each curve in Figure 11 corresponds to a given electricity spot price (all other parameters being equal).

Depending on the electricity price level, three cases can be distinguished. Either the slope of the profit function is positive and reaches positive values. In this case, the economic optimum is to operate the power plant at full load. A second case is when the slope is positive, but no positive value can be reached. In this case, operation at full load is still ensuring minimized losses. The duration for which the electricity prices remain at this level, as well as the start-up costs, define the economic trade-off at which the power plant should remain at full load or shutdown to restart later. The third case is when the profit function slope is negative. In this case, the load which minimizes the losses is the lowest load the power plant can reach before shutting down. Again, the electricity price profile and startup costs define the trade-off at which the plant should not park at minimum load but shutdown to start later. Thus, the trade-off between shutting down the plant and restarting it versus parking it at minimum load must be optimized, but this analysis demonstrates that there is no economic optimum in parking the plant at part-load above its minimum load. This option may, however, result from constraints like must-run conditions, power purchase agreements, or capacity reservations for frequency control.

4.2.2 Minimum up- and down-times

In the operations research and energy system analysis fields, the formulation of the unit commitment problem often includes minimum up- and down-times, to enforce a minimum duration between these events. As noted in [155] at page 199, minimum up- and downtimes often do not account for technical constraints but are introduced for risk or cost reduction purposes. Most of the dispatch optimization problem formulations, however, make use of minimum up- and down-times, whereas these result from a case-to-case economic trade-off in the event-based approach. This is made possible as the start-up and shutdown curves are explicitly accounted for. All technical restrictions can thus be included in these events description, and the operation or downtimes durations result from economic reasons only.

4.3 Intraday lead time

In this work, the intraday market is simulated as an arbitrage option. In case of high intraday prices and a power plant is in its "available" state, the arbitrage opportunity can only be taken if the price signal is forecasted long enough in advance for the power plant to start (typically 4 to 12 hours for a thermal power plant). The validity of the assumption depends on the evolution of the forecast quality with time. The longer the lead time (distance between the forecast point and the realization), the lesser the forecast quality and the higher the risk for the operator.

For all other situations, the power plant ramp-rate is the only limiting factor. For a power plant with 50 % minimum load and 2 %/min ramping capability (a rather non-flexible power plant), this correlates to about 25 minutes lead time between the dispatch modification and the forecasted intraday price realization. This corresponds approximately to the current lead time of intraday markets so that the assumption is valid for these cases.

4.4 Fleet effects

The single unit self-scheduling problem solved in this work considers a power plant owner dispatching one power plant, whereas, in reality, the power plant owner dispatches his power plant fleet. However, the first approach has been chosen, as the aim of this work is to provide a tool for the analysis of operational flexibility at individual generation technologies and eventually investment decision-making, which is more likely to concern one power plant than the whole fleet. When retrofitting more than one power plant of a fleet, optimizing the operation of the whole fleet might lead to a more realistic result regarding profit increase but does not allow attributing the benefits to the corresponding individual retrofits. Moreover, with the growing trend toward distributed energy generation, owners of vast fleets might become more seldom [156].

4.5 Stochasticity

Stochastic optimization allows accounting for risk and uncertainty, but with the main drawback to make the problem formulation more complex and the results less transparent [157]. The authors of [157] distinguish between problem formulations, which explicitly account for stochasticity and deterministic problem formulations using statistical inputs. For a review of the first ones see [157]. The approach proposed in this work does not include uncertainty within the problem formulation. Instead, uncertainty can be represented by a set of input scenarios with given probability distributions, thus requiring multiple optimization runs per scenario. Given the low computation time, this approach is acceptable. Moreover, scenario reducing methods can be applied, see [158, 159].

4.6 Section conclusion

The assumption's review has shown that the developed concepts are applicable in a liberalized market context with reasonable market liquidity for operational flexibility evaluations from the perspective of a price-taker merchant power plant operator. When comparing the simulated power plant dispatch and the real power plant dispatch, differences might occur due to non-applicability of the mentioned pre-requisites but also due to re-dispatch measures taken by TSOs, intraday or frequency control market designs not matching the design accounted for in this work or even other markets or agreements not accounted for in this work. The comparison of the simulated and real dispatch of the German Rostock power plant in December 2016, see Figure 17, however, demonstrates that both dispatches can be very close. The real dispatch is taken from the entsoe Transparency Platform [160].

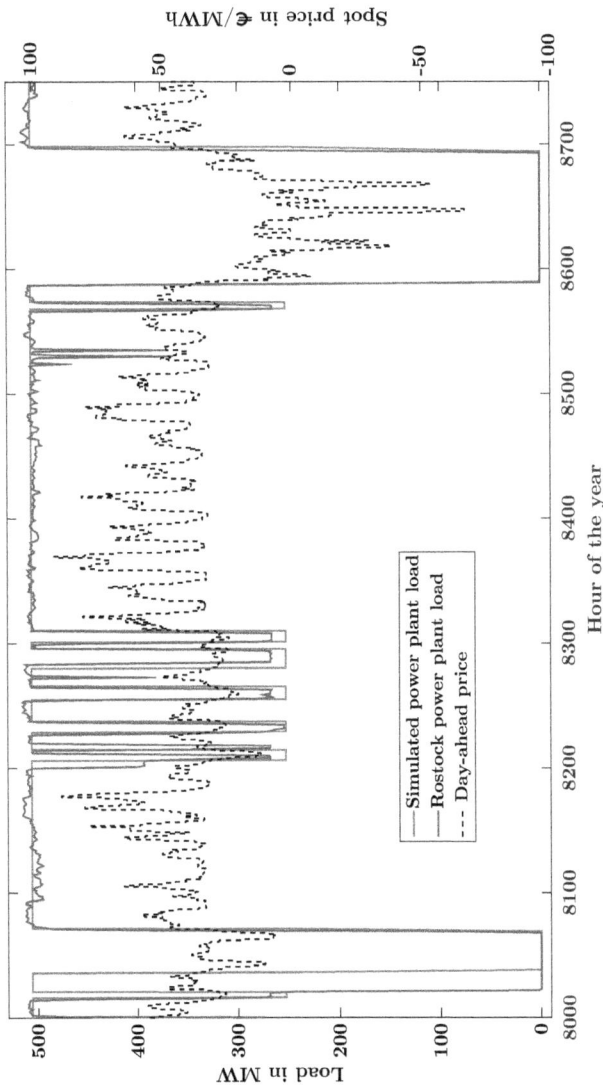

Figure 17 Comparison of the real and simulated dispatch of the Rostock power plant in December 2016. The real power plant dispatch is taken from [157]. The simulated dispatch is the result of a dispatch optimization using inputs presented in the case studies of Part C.

5   Part conclusion

The work performed in this part provides the methodological fundament to solve the single unit self-scheduling unit commitment and economic dispatch problem to perform the quantitative long-term high-resolution techno-economic assessment of operational flexibility at individual power plants. A power plant description suited for the analysis of operational flexibility has been developed and included in the uniform time discrete state-of-the-art mixed integer formulation of the self-scheduling problem from a price-taker perspective. A sensitivity analysis has shown the influence of the power plant description on the value of flexibility in monetary terms. This power plant description has also been formulated in the formalism of a finite state machine, introduced for the proposed event-based approach to the self-scheduling problem. This approach avoids uniform time discretization, thus leading to better solution quality. To provide a comprehensive operational flexibility analysis, a review of the relevant markets has been proposed. A case study has shown that a flexibility study should include at least the frequency control and intraday markets. As the literature review has shown, including these markets is not a straightforward task. In this work, it has been proposed to address the capacity reservation for frequency control as an opportunity cost calculation and the intraday market as an arbitrage opportunity. The case study has shown that, despite the intuition that capacity reservation for frequency control might reduce the value of a flexibility improvement compared with the day-ahead market alone, the analyzed flexibility improvement has the highest value in this market environment. This aspect will be further analyzed in the case studies of Part C.

Part C

Case studies

1 Introduction

What is the value of conventional power plants' operational flexibility? What are good technical solutions for thermal power plants' flexibility? What is a proper flexibility retrofit from a commercial point of view? Are the markets well designed for flexibility incentives? All these questions require to quantitatively assess the value of operational flexibility at thermal power plants in their market environment. This chapter thus aims at demonstrating how the tools developed in this work can be used to answer these questions at selected power plants and in selected market environments. Different types of case studies are performed, which require defining market environments, power plants, and operational flexibility improvements. Calculations are conducted, which aim at quantifying the value of a given retrofit as a function of the market context and power plant technologies. These support investment decisions, from a power plant operator's perspective as well as from an equipment manufacturer's perspective. The performed calculations are also used to assess further aspects like the quantification of the arbitrage possibilities in the intraday market and frequency control market and thus to determine whether these markets are "well designed" for flexibility incentives.

As noted in [161], technical solutions are available to make coal-fired as well as gas-fired power plants more flexible. The benefit of these flexibility improvements to the system depends on the nature of the said energy system. The authors of [161] suggest that coal retrofits are more beneficial to a coal-dominated system with low coal prices than to a gas-dominated system with low gas prices. Power plants with gas turbines (simple and combined cycle) are reported to have better ramp-rate capabilities than those with steam turbines, see, e.g. [15]. The coal-fired power plant fleet, an aging fleet, might thus be a better candidate for flexibility improvements. The same study also reports better part-load efficiencies at gas turbine power plants, but the minimum load capabilities are better at coal-fired power plants than gas-fired ones. A fair analysis would thus compare the efficiency drop for a comparable load drop. Regarding emissions, [15] reports that on average, at minimum complaint load, gas turbines produce more NOx but less CO_2 and SOx emissions than coal-fired power plants. In Europe, the current fuel prices lead to a merit order (ranking the generation capacities by increasing marginal operation costs) with the gas-fired generation technologies at the most expensive end. Depending on the countries' specificities, the cheapest generation technologies after renewables, and thus the ones primarily providing the residual load, are coal and/or nuclear based. The social question of acceptance is not part of this work, but it is worth noting that the role of coal is a controversial topic, with, e.g., the coal exit planned in France, the drastic reduction of coal generation in the United Kingdom and the debate of a coal exit in Germany. On the other hand, some countries like Poland, India, and China are likely to rely on coal to bridge the times of the energy transition toward a sustainable system. The case studies performed in this work, despite being applicable to all dispatchable technologies, focus on answering the research questions for coal-fired generation technologies.

2 Case study catalog

2.1 Power plant selection

2.1.1 Type cluster and data collection

A straightforward clustering of thermal coal-based generating plants makes use of the combustion technology and characteristics of the steam cycle. The combustion technologies are divided into the fluidized bed technology, pulverized coal technology, and stoker technology. Due to the predominance and more extended operation experience [162], this study is limited to pulverized coal technologies. The steam cycle is often subdivided along the growing steam pressure and temperature in subcritical (UC), supercritical (SC) and ultra-supercritical cycles (USC), with even advanced ultra-supercritical cycles (aUSC). This wording refers to the critical point of water so that the steam conditions are subcritical below 221 bar and ultra-supercritical above 221 bar and 590 °C. All but the last will be addressed in this study. This section details the literature review performed to collect data for power plant model parameters. The comprehensive data set and power plant parameters can be found in Appendix.

The data has been collected from various, also grey, literature resources. The general data regarding power plant type, nominal net output, efficiency and part-load behavior are taken from [163], [164], [165], [166], [167]. The data set is thus based on real-world power plants, but missing/undisclosed data is inferred from other literature resources. The timely variation of the power plant efficiency with ambient temperature is not taken into account in the power plant model. When simulating power plants in a warmer environment, as for the case study in India or Turkey, the reduced plant efficiency is however taken into account [168]. When information is missing about the minimum load efficiency, assumptions are made based on [165], [169] and [167]. Figure 1 in [165] provides a data basis for part-load efficiency at subcritical hard coal-fired units. At 30% of nominal load, the loss of efficiency lies by 10 %. Based on [169], supercritical units are assumed to have only the half of this loss. [167] shows a factor two between lignite and hard coal-fired units part-load degradation of efficiency. Figure 18 compares the part-load efficiency degradation of the defined power plants. [164] provides general data about minimum load levels. Nominal load efficiencies, when not available, are taken from [163]. A rate of 8% house load is assumed for auxiliary equipment. Data from the literature regarding the start-up fuel consumption has been compared to non-disclosable power plant operator data, see Table 16. The literature resource [124] matches the mean values collected from plant operators and is therefore used for the case study. The repartition between main fuel and ignition fuel observed in the plant operator data is used in the case study. Based on [170], lignite and hard coal-fired plants with the same steam cycle characteristics are supposed to require the same energy input. Supercritical units are said in [169] and [124] to require more heat to start-up than subcritical units, which is also assumed here. [164] on the contrary, states that sliding pressure operated plants (commonly also supercritical plants) require a lower overall heat input. Start-up durations are collected from [171], [164] and [172]. The latter also provides a basis for the start-up electricity production curve. The unit ramp-rates are taken from [164], [166], [167], and [173]. As no specific data is provided, the power plant ramping capability for secondary and tertiary frequency control is assumed to be the same as the load following ramp-rates, and the standard 5 % per 30 seconds for primary control is assumed. The ramp-rates after a start-up event are assumed to be decreasing with the component temperature: the colder the start-up, the slower the ramp-rate. The event-specific maintenance costs, see Table 15, are taken from [174], [175] and [124]. The range of values found in the literature is vast. According to [174], supercritical and ultra-supercritical plants have more considerable maintenance costs than subcritical plants. [176] indicates that supercritical units see a 25 % cost increase compared to subcritical units. However, older plants have more considerable cycling-related maintenance costs than newer ones, especially when they have reached half of their lifetime [124]. As the subcritical fleet is older, a comparable event-specific maintenance cost is assumed for subcritical and supercritical technologies. The static continuous maintenance costs component introduced in this work's power plant model adds up over the whole plant lifetime, independent of its operation and comprises aging of components like creep, consumables like ammonium for flue gas treatment as well as crew costs. In the operation regime above minimum technical load, a load proportional part comprises aging of components like creep due to pressure or centrifugal forces whereas the static component is increased due to higher crew or consumables costs. The literature references do not distinguish between the operating and available operating regime maintenance costs cm_0 and cm_h, so that this work assumes that 40 % of the costs found in [177] and [176] are allocated to the static continuous component. The load proportional maintenance cost for baseload operation as well as the load following costs, are taken from [124], see Table 17.

Combined heat and power (CHP) plants are often distinguished between those driven by the electricity demand or those driven by the heat demand. This work assumes that the heat demand has priority (must-run condition when heat is required) and that the electricity production is optimized within the remaining capacity and flexibility limits. In Denmark, the district heating market is regulated. For large scale CHP plants, the electricity production follows the deregulated market rules and the heat the regulated market rules. The non-profit principle applies so that heat is remunerated at production cost [178]. It is assumed here that these production costs approximately equal the fuel costs. The difficulty then lies in the allocation of fuel consumption between electricity and heat. For tax purposes, the Danish Energy Agency recommends the use of a 120 % efficiency on the heat side, so that 0.83 MWh fuel is allocated to a heat production of 1 MWh [178]. This recommendation is used to define the heat remuneration when simulating the Danish market. For Germany, no such rule or publicly available data is provided so that the assumption is made that the prices are the same as in Denmark.

2.1.2 Power plant identification and flexibility measure

The data collection reported above conducted to the definition of a set of power plants, which can be found in detail in Appendix.

The naming code used for the power plant identification contains three tags. The first one allows distinguishing between subcritical units (UC), supercritical units (SC), and ultra-supercritical units (USC). The second part of the code indicates whether the plant is hard-coal (H) or lignite (L) fired. The third part of the code is used to distinguish between power plants with same fuel and steam cycle parameters. It is either based on a real power plant with similar characteristics or a particular characteristic of the plant (older or smaller plant for instance).

This section provides a comparison of the technical capabilities of the power plants, with a particular mind to operational flexibility. Figure 18 compares the part-load efficiency degradation of the defined power plants. The load and efficiency are normalized to allow for comparison, and the efficiency is depicted until 10 % of the nominal load, even if the power plants do not reach this low level of minimum stable operation. As the literature review has shown, operational flexibility metrics define a flexibility trinity composed of ramp-rate, power and energy. Ramp-rates refer to the load difference between two steady states divided by the time required for the power plant to change the load from the first level to the second one. Ramp-rates are thus the parameter allowing to derive energy from power. The power plant model introduced in this work distinguishes between the ramp-rates after each start-up type, before shutdown and the load-following ones. These ramp-rates are considered constant over the whole minimum load to full load range. The minimum complaint load, which is the lower bound to power, is thus depicted with the ramp-rates, see Figure 19. With regard to ramp-rates and minimum load, the most flexible plants are thus located in the upper left side of the figure, whereas the less flexible ones are located in the right hand side lowest part of the figure. The power plant efficiency gives an indication of the power plant's relative marginal operating costs. As the start-ups are costly events, these events' durations are clustered and depicted with the related costs.

Table 15 Literature review of start-up maintenance costs

Plant type	Coldest start maintenance costs	Warm start maintenance costs	Hottest start maintenance costs
Undifferentiated	90000€/event [174] 48678€/event [175]	5400€/event [174]; 29207€/event [175]	4600€/event [174]; 9736€/event [175]; 26000€/event [124]
Small subcritical plants	147€/MW[124]	157€/MW[124]	94€/MW[124]
Large subcritical plants	105€/MW[124]	65€/MW[124]	59€/MW[124]
Supercritical plants	104€/MW[124]	64€/MW[124]	54€/MW[124]

Table 16 Literature review of start-up fuel consumption

Plant ID	Coldest start specific fuel consumption in MW_{th}/MW	Warm start specific fuel consumption in MW_{th}/MW	Hottest start specific fuel consumption in MW_{th}/MW
Hard coal- fired subcritical plants	3.4[124]	2.4[124]	1.8[124]
Hard coal-fired supercritical plants	0.9[179]; 5.9[124]	5[124]	0.6[179]; 2.96[124]

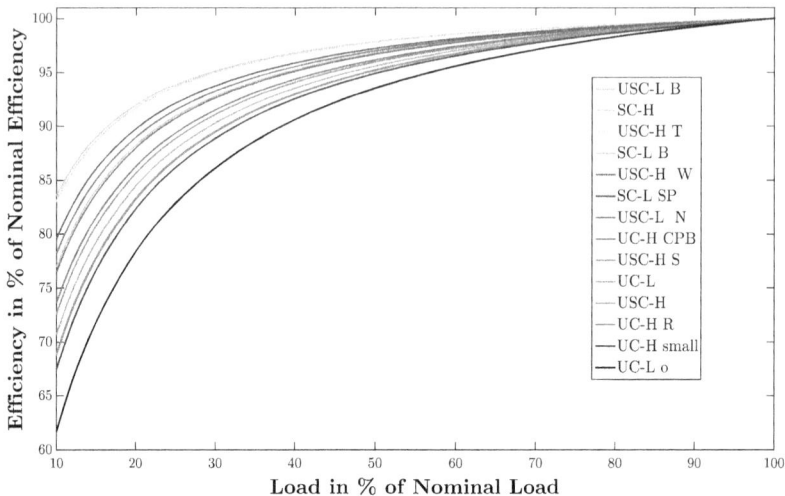

Figure 18 Power plant set's efficiency degradation at part-load. The load and efficiency are normalized to allow for comparison.

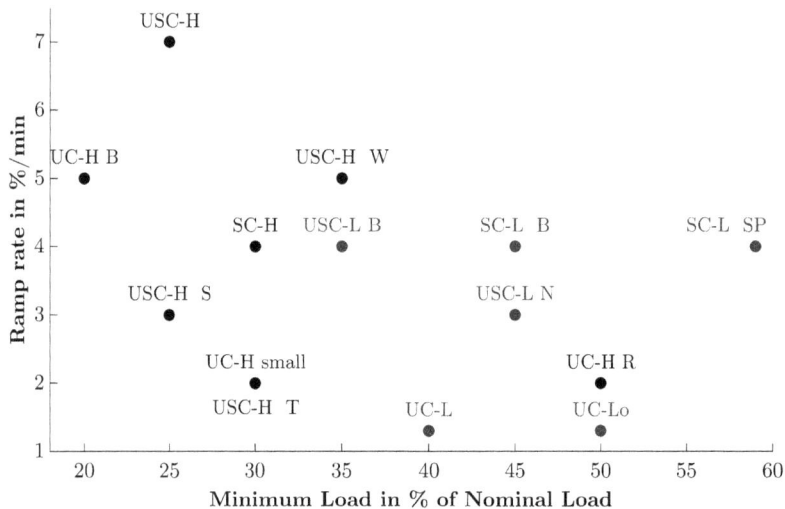

Figure 19 Ranking of the selected power plants according to their ramp-rate and minimum complaint load

Table 17 Literature review of maintenance costs and assumptions

Plant	cm_0	cm_h	cm_1	Load follow
Undifferentiated	3.2€/MW/hour[177]; 2.6€/MW/hour [176]	N.A.	N.A.	N.A.
Small hard coal-fired subcritical plants	1.04€/MW/hour	1.56€/MW/hour	2.82€/MWh [124]	3.34€/MW [124]
Large hard coal-fired subcritical plants	1.2€/MW/hour	1.8€/MW/hour	2.68€/MWh [124]	2.45€/MW [124]
Hard coal-fired supercritical plants	1.3€/MW/hour [180]	1.95€/MW/hour	2.96€/MWh [124]	1.96€/MW [124]

2.2 Operational flexibility options

The previously selected power plant set offers a wide range of power plant flexibility. This section further provides a set of flexibility improvements. The first subsection describes a flexibility improvement to be applied to the whole power plant set. The second subsection's focus is set on two German power plants. The focus of this section is not on the description of the technical flexibility improvement in itself, as it has been done in the literature, but rather on its impact on the power plant's operational flexibility. Data available in the literature is collected and confronted and used to support the assumption selection.

2.2.1 Decoupling the power plant dynamics

Indirect firing is a form of fuel storage to decouple the power plant's steam cycle dynamics from the mill dynamics. Coal is milled, dried and eventually stored ahead of the combustion process [164]. This improvement allows for an increase in the power plant ramp-rates and reduction of start-up durations and ignition fuel consumption [164]. The increased lifetime consumption due to the steeper ramp-rates is documented in (Kumar et al. 2012). This increase depends on the past operation of the power plant and the related accumulation of creep and fatigue. As the power plant is assumed to go through a retrofit improving its ramping capability, this work assumes slightly lower multiplying factors than in (Kumar et al. 2012).

Table 19 shows the data collection for the ramp-rate increase reached via indirect firing and the selected data assumption for the different technologies. The multiplying factor for the maintenance costs increase is included. Table 20 and Table 21 show the data collection for the other improved power plant parameters at hard coal- and lignite-fired power plants respectively. The last column displays the assumption made for the case study. The start-up process of the retrofitted power plants will follow the same process as the baseline power plants, but end at the retrofitted lower minimum load, thus reaching a start-up duration reduction in line with the power plant's particular process. For the hard coal-fired power plants, the indirect firing allows reducing the minimum firing rate to 10% of its nominal value. The firing rate being proportional to the load [51], the documents reporting the performances of indirect firing assume a minimum load reduction to also 10% of its nominal value. Operational experience is not reported, but commercial operation at 12% of nominal load is reported at the German power plants Heilbronn 7 and Bexbach [51]. This performance is achieved via one-mill operation but shows that very low load operation at these levels is also possible on the steam cycle side. For lignite-fired power plants, the use of dry lignite does not allow such deep load operation, as combustion stability is a bigger issue at lignite-fired power plants [51], but still reduces the achievable minimum low by 50% as reported in Table 21.

The investment costs are estimated to 50 Mio € plus 30 % of new power plant cost in [181]. A Vattenfall flyer advertises an investment of 13 Mio€ for the use of dry lignite at its Jänschwalde lignite-fired power plant.

Table 18 Data collection and assumption for ramp-rate increase related maintenance costs

	Load follow maintenance costs multiplying factor for faster ramp-rates	Selected load follow maintenance costs multiplying factor for faster ramp-rates
Small subcritical plants	2 to 8 [124]	1.5
Large subcritical plants	1.5 to 10 [124]	1.2
Supercritical plants	1.5 to 10 [124]	1.2
Ultra-supercritical plants	-	1.2

Table 19 Ramp-rate increase and related maintenance cost data assumption

	Baseline	Retrofit
Ramp-rates according to [163]	3%	6% - 10%
Technology	Selected ramp-rate after indirect firing retrofit	Selected load follow maintenance costs multiplying factor for faster ramp-rates
Lignite-fired subcritical	4%/min	1.5
Hard coal-fired subcritical	6%/min	1.2
Supercritical plants	6%/min	1.2
Ultra-supercritical plants	6%/min	1.2

Table 20 Data collection and assumption for indirect firing at hard coal-fired power plants

	Baseline	Retrofit	Selected retrofit
Minimum load as % of nominal load	25-30% [164] [166]; 40% [182];	<10% [164] [166]; 25% [182] 10% with burner retrofit [182];	10%
Ignition fuel demand as percent of baseline consumption	100% [164]	5% [164]; 10% [182]	10%
Efficiency at full load	46% [166]	N.A.	No modification
Ramp-rate	4%/min [164]; 2-5%/min [164]	7%/min [164]; 10%/min [164]	See Table 19
Cold Start-up duration	300-480min [166]	300-480min [166]	Duration defined by the minimum load reduction and baseline curve
Hot Start-up duration	140min [166]	120min [166]	Duration defined by the minimum load reduction and baseline curve

Table 21 Data collection and assumption for indirect firing at lignite-fired power plants

	Baseline	Retrofit	Selected retrofit
Minimum load as % of nominal load	40% [166]	20% [166]	20%
Ignition fuel demand as percent of baseline consumption	N.A.	N.A.	10%
Efficiency at full load	43.5%[166]	N.A.	No modification
Ramp-rate	3%/min [166]	4%/min [166]	See Table 19
Cold Start-up duration	380-480min [166]	300-480min [166]	Duration defined by the min. load reduction and baseline curve
Hot Start-up duration	66-140min [166]	120min [166]	Duration defined by the min. load reduction and baseline curve

2.2.2 Flexibilization of power plant Schwarze Pumpe and Rostock

The power plant Rostock is a subcritical hard coal-fired power plant with fuel oil as a support fuel. The Schwarze Pumpe power plant is a supercritical lignite-fired power plant with fuel oil support. Published data about these power plants have been used to define the power plants UC-H R and SC-L SP in the previous section 2.1. Both power plants have been cited in some studies regarding power plant retrofits. The study [167] addresses a retrofit at the Rostock power plant, which decreases the minimum load and increases minimum load efficiency and ramp-rates, see Table 22. In the work at hand, the ramp-rate increase is modeled with an increase in maintenance cost by a factor 1.2 [124]. The retrofit at the Schwarze Pumpe power plant addressed in [173] decreases the minimum load from 59 % to 25 % of the nominal load [173]. The study [51] lists possible retrofits leading to power plant flexibilization. The minimum load reduction, for instance, might be reached via one-mill operation or dried-lignite firing. The ramp-rate increase might be achieved using thick-wall components retrofitting or upgrading the power plant control. Better part-load efficiency might be achieved using variable speed drives for pumps, allowing for operation at design load [164].

Table 22 Data collection: Power plant Rostock, data taken from [167]. SCR: selective catalytic reduction, P: nominal load.

Rostock	Baseline	Retrofit
Minimum load as % of nominal load	50%	35%;
		25% (requires retrofit for SCR)
Efficiency at minimum load	40.6%	@37%P: 39.5%,
		@30%P: 37.5%
Ramp-rate	2%/min	4%/min

2.3 Country and market selection

2.3.1 Market selection rationale

The International Energy Agency (iea) defines four levels of variable renewables integration [183]. The more advanced phase, the Phase 4, is about short-term stability. Phase 3 is titled "Flexibility is key." Phase 2 requires solely better operations to accommodate variable renewables. At Phase 1, variable renewable energy (VRE) is not noticeable to the system operator. For each of these levels, a selection of countries is provided, which has been used to select

(based on data availability, as time series of electricity spot prices
are required) the countries assessed in this case study, namely
Denmark, Ireland, Germany, Spain, India, Sweden, France, and
Turkey. The two last countries have not been classified in [183] but
are assumed to belong to the Phase 1 category. Among the selected
countries, fossil-fuel based electricity generation plays very different roles, even within a given category, as the example of Sweden
and India shows. In fact, coal, oil, and natural gas are relied on for
only 2.4 % of the electricity generation in Sweden (see iea country
profile) whereas in India these fossil fuels are relied on for 82 % of
the electricity production (see iea country statistics). The applicability of the developed methodology to the Indian context is however questionable, as this market is not fully liberalized [184]. As
operational flexibility, and especially lower minimum load is a
much-demanded topic in India, the case study will be performed
using the electricity spot prices at the Indian energy exchange. As
in Sweden, the reliance on fossil fuels is scarce in Ireland. Therefore, the case study calculations are limited to the countries Denmark, Germany, Spain, India, France, and Turkey. The share of
variable renewable energy and some remarks to the countries' specificities are to be found in Table 23.

2.3.2 Parametric frequency control market

Table 24 summarizes the average price ranking for balancing found
in [151] as well as the capacity procurement scheme taken from
[185]. The definitions provided by the European Network of
Transmission System Operators for Electricity (entsoe) are the
following: a bilateral market relies on contracting between TSOs
and grid users, an organized market relies on a voluntary
participation of grid users on a market platform, mandatory offers
are the obligatory offering of remaining capacity and mandatory
provision is the obligatory reservation of a given capacity.

In France, automatic frequency restoration (aFRR) and frequency
containment (FCR) are reserved via obligations and the price for
capacity regulated at about 18 €/MW/h. Both reserves can be
exchanged over-the-counter [186]. Manual frequency restoration reserves are divided into rapid reserves (RR) and complementary reserves (RC). These reserves can either be contracted for the entire
week or for peak/off-peak times of the week via an annual tender
call [187]. Denmark is divided into two zones; DK1 (West, connected to Germany) and DK2 (East, connected to Sweden). DK1 has
aFCR, aFFR supply ability, aFRR, and mFRR. DK2 has aFCR-
Disturbances, aFCR- Normal operation, aFRR supply ability and
mFRR [188]. Primary reserves (aFCR) in DK2 are organized in col-

laboration with Sweden. Tertiary reserve (mFRR) procurement is common to Norway, Sweden, Denmark, and Finland. Germany has four transmission operators sharing a common platform for the procurement of three ancillary services: aFCR, aFRR, and mFRR. Spain has secondary control (aFRR), tertiary control (mFRR), and additional upward reserve power. All these services are optional, but if subscribed to the tertiary control, bids are mandatory, and energy only is remunerated [189]. Primary control is a mandatory non paid service [190]. The parametric frequency control market model introduced within this work (Part B section 3.2) allows representing all these markets within the same framework. The data collection used for the calculation is shown in Table 28. Data for Tukey is included despite the mandatory provision scheme. The price signals are the weighted average of realized prices as detailed in section 3.2.2, Part B.

Table 23 Summary of some countries' characteristics. VRE: variable renewable energy.

VRE integration phase[183]	Country	Electricity generation 2016	Remarks
4	Denmark	29% coal, 45% VRE	Majority of thermal power plants are CHP plants. Has two price zones, DK1 and DK2.
4	Ireland	17% coal, 22% VRE	Mandatory gross pool with ex-post price formation. Different from other European countries which use self-dispatch and define prices ex-ante. Only two coal-fired power plants. Capacity payments.
3	Germany	43% coal, 18% VRE	High reliance on coal and renewables. Single price zone and four control zones.
3	Spain	14% coal, 23% VRE	Capacity mechanism.
3	Portugal	22% coal, 24% VRE	The Spanish and Portuguese electricity markets are so well integrated, that the spot market prices converge.
2	India	75% coal, 3.5% VRE (year 2015)	Almost only subcritical power plants due to coal quality. The main drawback is the high ash content of coal. Coal wash would increase fuel costs. Efficiency decrease due to ambient temperatures.
2	Sweden	1%coal, 10% VRE	Four price areas. Only 1% of electricity generation is based on coal.
1	France	2% coal, 6% VRE	France to close its last 4 coal power plants in the next 5 years. Decentralised capacity obligation from 2017 on.
1	Turkey	34% coal, 6.6% VRE	Efficiency decrease due to ambient temperatures. Intraday market introduced in 2015.

Table 24 Frequency control market schemes in selected countries

Countries	Price ranking [151] (1 =highest)	Capacity procurement scheme mFRR	Capacity procurement scheme aFRR	Capacity procurement scheme FCR
Ireland	4	N.A.	N.A.	N.A.
Denmark	5	Organised market	Bilateral market	Organised market
Germany	6	Organised market	Organised market	Organised market
Spain/Portugal	1	Mandatory offers	Organised market	Mandatory provision
India	-	N.A.	N.A.	N.A.
Sweden	2	Organised market	Organised market	Organised market
Turkey	-	N.A.	Mandatory provision of unused capacity, not remunerated	Mandatory provision
France	3	Organised market	Mandatory provision	Mandatory provision

2.3.3 Fuel and emission costs

The German statistics office [191] collects the monthly variation of all fuel prices for Germany with 2010 as basis year. The prices for the year 2010 are taken from [170]. The mean value of this data is summarized in Table 25, but the calculations performed for this case study make use of the monthly variations of [191]. Spain and France are assumed to have the same fuel prices as Germany. The fuels prices for Denmark can be derived from the Danish statistics database [192]. The statistics ENE1HA and ENE4HA for electricity production indicate the amount used yearly for each fuel and the corresponding monetary value. Lignite is not used for electricity production in Denmark. The values are summarized in Table 26. Indian thermal coal used for electricity generation is mainly relying on domestic coal and imports from Indonesia [193]. Domestic coal prices for power plants are notified by the Coal India Limited; a government of India enterprise. Resulting assumptions are presented in Table 26. Monthly prices for oil and natural gas are reported by the ministry of petroleum and natural gas [194] and used for the simulations. The Turkish price assumptions are also summarized in in Table 26.

To support sustainability, a possible mechanism is the pricing of greenhouse gas emissions. As reported by the World Bank, over 40 nations and 27 subnational jurisdictions worldwide were pricing 15 % of the global CO_2 emissions in 2017 [195]. These pricings are distinguished between carbon taxes and cap-and-trade systems.

The World Bank defines carbon pricing as follows: "carbon pricing refers to initiatives that put an explicit price on greenhouse gas emissions, i.e., a price expressed as a value per ton of carbon dioxide equivalent (tCO_2e). These initiatives include not only emission trading systems (ETS), carbon taxes, offset mechanisms, and results-based climate finance, but also internal carbon prices set by organizations. Policies that put an implicit price on carbon, for example, removal of fossil fuel subsidies, fuel taxation, support for renewable energy, and energy efficiency certificate trading, are not included [...]." The OECD definition of taxes is as follows: "taxes include carbon taxes and- importantly- specific taxes on energy use generally." It is, therefore, necessary to distinguish between explicit and implicit carbon taxes. The first ones are a price directly based on carbon (that is, price per tCO_2e). The second ones need to be converted in a pricing per ton of CO_2e and thus to know or assume the carbon content of the taxed good. It is made use of [195] [196], [197] and [198] to collect the information summarized in Table 27. The International Energy Agency's energy prices and taxes country notes [199] provide the following information: Ireland has an oil tax with a carbon component introduced in 2009. Natural gas and coal for electricity generation are exempted from the carbon tax. Denmark has a CO_2 tax additional to natural gas and coal taxes. Power plants for electricity production are exempted from these fuel taxes, whereas combined heat and power (CHP) plants are not. In Germany, coal used for electricity generation is exempted from excise duty. Spain has a tax on coal, from which coking coal is exempted. No explicit carbon tax is in place. Portugal introduced a carbon tax in 2015 as a component of the tax on oil and energy products. Sweden has an energy tax on fossil fuels and electricity as well as a CO_2 tax, which does not apply to electricity and uses covered by the European trading system. In Turkey, steam coal consumption is not subject to excise taxes. France taxes all fossil fuels with an extra carbon component introduced in 2014. Power plants, including CHP plants, are exempted from the tax on natural gas. All these countries have taxes on electricity consumption.

Table 25 Data collected for European fuel prices

	Germany, France, Spain		
	Year 2010	Year 2014	Year 2016
Lignite in €/MWh	1.5 [170]	1.13*1.5 [191]	1.107*1.5 [191]
Hard coal in €/MWh	10.4 [170]	0.857*10.4 [191]	0.805*10.4 [191]
Natural gas in €/MWh	21 [170]	1.139*21 [191]	0.878*21 [191]
Oil in €/MWh	38.3 [170]	1.222*38.3 [191]	0.59*38.3 [191]
CO_2 certificate in €/ton	13 [170]	6	3.91

Table 26 Data assumption for Turkish, Indian and Danish fuel prices

	Turkey		India		Denmark	
	Year 2014	Year 2016	Year 2014	Year 2016	Year 2014	Year 2016
Lignite in €/MWh	2	1.9	1.99	2.16	-	-
Hard coal in €/MWh	6	5.68	7.89	5.59	9	8.7
Natural gas in €/MWh	18	18.2	17.2	10.6	31.5	29.3
Oil in €/MWh	46.83	23	57.3	25	56.5	22

Table 27 Carbon pricing data collection

	Explicit Carbon tax [195] [196]	Explicit carbon tax for electricity production [197]	Explicit and implicit carbon tax on electricity [198]	Emission trading system
Ireland	24USD/tCO_2e	Power stations exempted	2.3EUR/tCO_2e	EU ETS
Denmark	27USD/tCO_2e	CHP only	104.6EUR/tCO_2e	EU ETS
Germany	None	None	26.2EUR/tCO_2e (no explicit carbon tax, tax on energy use)	EU ETS
Spain	None	None	2.5EUR/tCO_2e (no explicit carbon tax, tax on energy use)	EU ETS
Portugal	8USD/tCO_2e	Power stations exempted	1.4EUR/tCO_2e (carbon tax introduced in 2015 not included)	EU ETS
India	None	None	0.6EUR/tCO_2e (no explicit carbon tax, tax on energy use)	None
Sweden	140USD/tCO_2e	Power stations exempted	193.1EUR/tCO_2e	EU ETS
Turkey	None	None	11.5EUR/tCO_2e (no explicit carbon tax, tax on energy use)	None
France	36USD/tCO_2e	None (carbon tax is on fuels not covered by ETS)	12.4EUR/tCO_2e	EU ETS

Table 28 Country-specific data for the parametric frequency control market model

	Program unit size	Minimum bid size	Full activation time	Peak and off-peak product	Symmetrical product
Denmark DK1 FCR	4 hours blocks daily	0.3MW	30 seconds	No	No
Denmark DK1 aFRR supply ability (2015)	Month	1MW	15 min	No	Yes
Denmark DK2 aFRR supply ability (2015)	Month	1MW	5min	No	Yes
Denmark DK1 aFRR (LFC)	Less than a month	1MW	15 min	No	yes
Denmark DK2 FCR-N with Sweden	Hours of a day	0.3MW	150 seconds	No	Yes
Denmark DK2 FCR-D with Sweden	Hours of a day	0.3MW	25 seconds	No	Up only
Denmark DK1+DK2 mFRR	Hours of a day	10MW	15 min	No	No
Germany aFCR	Week	1MW	30 seconds	No	Yes
Germany aFRR	Week	5MW	5 min	Yes	No
Germany mFRR	Day, 6 blocks of 4 hours	5MW	15 min	No	No
Spain/Portugal aFRR	Hour	10MW	5 min	No	Yes
Spain/Portugal mFRR	Hour	10MW	15 min	No	No
Sweden aFCR	Hours of a day	0.3MW	150 seconds	No	Yes
Sweden mFRR	Hours of a day	10MW	15 min	No	No
Turkey aFCR	Hour	1% of capacity	30 seconds	No	No
Turkey aFRR	Hour	Contracted	5min	No	No
France aFCR	Hour	1MW	30 seconds	No	Yes
France aFRR	Hour	1MW	<15 min	No	Yes
France mFRR	Week with peak and off-peak periods	10MW	13 min RR 30 min RC	Yes	Up only

3   Case study results

3.1 Market analysis

The data collection provides for market environments and power plants with differentiated characteristics, which are analyzed in the case studies. This section's focus is on the design of the markets and the incentives sent regarding flexible operation.

3.1.1 Growing shares of renewables-based electricity production

The retroactive observation of the power plant's dispatch in the selected markets is used to analyze the effect of growing shares of variable renewable-based electricity generation (VRE). This observation compares the years 2014 and 2016, for which the share of VRE is to be found in Table 29. The dispatch of each power plant defined in section 2.1 is optimized in the day-ahead market of each of the selected countries, once for the year 2014 and once for the year 2016. The comparison is made based on the yearly profit reached by each power plant in its market environment, and the number of load following and start-up events it operates in a year. The yearly profit and cycling events are thus the chosen metrics for this analysis. The obtained values are analyzed using boxplots, see Figure 20. The central mark indicates the median, the value separating half of the data sample. The bottom edge of the box indicates the 25th percentile, the value below which 25 % of the data sample is to be found. The top edge of the box indicates the 75th percentile, the value below which 75 % of the data sample is located. The top whisker indicates the maximum, whereas the bottom whisker indicates the minimum of the data sample. Red crosses mark outliers. Outliers are defined as the values which are outside of the range defined by adding (subtracting) one and a half times the box size to the 75th (respectively 25th) percentile. The results show that in all market environments apart from France, the generated yearly profit tends to decrease in 2016 compared to 2014. At the same time, the amount of part-load events remains stable or slightly decreasing except for Turkey. The year 2016 in the Turkish market environment led to a sharp increase in cycling, concomitant to a sharp increase in renewables-based electricity generation. It is noticeable, that among the number of cycling-events, the part-load operation dominates. The comparatively low number of start-up events operated by the hard-coal fired power plants compared to other publications (see e.g. [84]) is due to the maintenance costs accounted for in our model. The cost estimations accounting for start-up specific lifetime consumption used in this analysis lead to high start-up costs. A hard-coal fired power plant with 3.5 MWhth/MW start-up hard-coal consumption and 2.4 MWhth/MW start-up natural gas consumption and 104€/MW maintenance costs lead to 180 €/MW start-up costs, assuming an 8 €/MWhth hard coal price and a 20 €/MWhth natural gas price. For such an 800MW power plant, the start-up costs would lie by 144 000 € per start (first ignition until minimum load).

The number of start-ups in the year 2016 compared to the year 2014 shows an increase in all market environments except for France and Spain, where it decreases. The amount of starts remains stable in India. The number of cycling events in the Indian context is almost the same for all power plants in the data set. The lower variability of the number of cycling events compared to the other market environments might be a result of the low participation to the electricity spot market and the not fully liberalized market context. The electricity spot prices do not reflect the variability and needs of the system and might thus not be an appropriate input for a self-scheduling scheme. Despite the slight increase in VRE production in France, it seems to be below a limit for which the power plants would be incented to cycle more. As the amount of VRE production remains stable in Spain/Portugal and India, so does the amount of cycling events. The yearly profit remains stable in Spain, whereas it decreases in India. The steepest increase of VRE happens in Turkey, and the amount of cycling events grows accordingly, accompanied by a decrease in yearly profit. The same is observed to a lesser extent in the German and Danish market environment. This indicates that these day-ahead price signals incent the power plants to operate more flexibly, but do not compensate for the profit difference induced by this more flexible, system-friendly dispatch.

The next sections thus analyze whether the intraday and frequency control markets offer an additional source of revenues and how these impact the flexibility of the power plant dispatches. The performed case studies are based on the optimized dispatch of the set of power plants in all market environments and on the comparison of the obtained dispatches with and without the additional markets. For each of these markets, first frequency control and then intraday, the results are first commented at a selected power plant and then at the full set of power plants.

Table 29 Comparison of the share of variable renewable energy for electricity production in the years 2014 and 2016 in selected countries

Country	Electricity generation 2014	Electricity generation 2016
Denmark	42.5% VRE	45% VRE
Germany	14.9% VRE	18% VRE
Spain	23.6% VRE	23% VRE
Portugal	24.1% VRE	24% VRE
India	3.3% VRE	3.5% VRE (year 2015)
France	4.1% VRE	6% VRE
Turkey	3.4% VRE	6.6% VRE

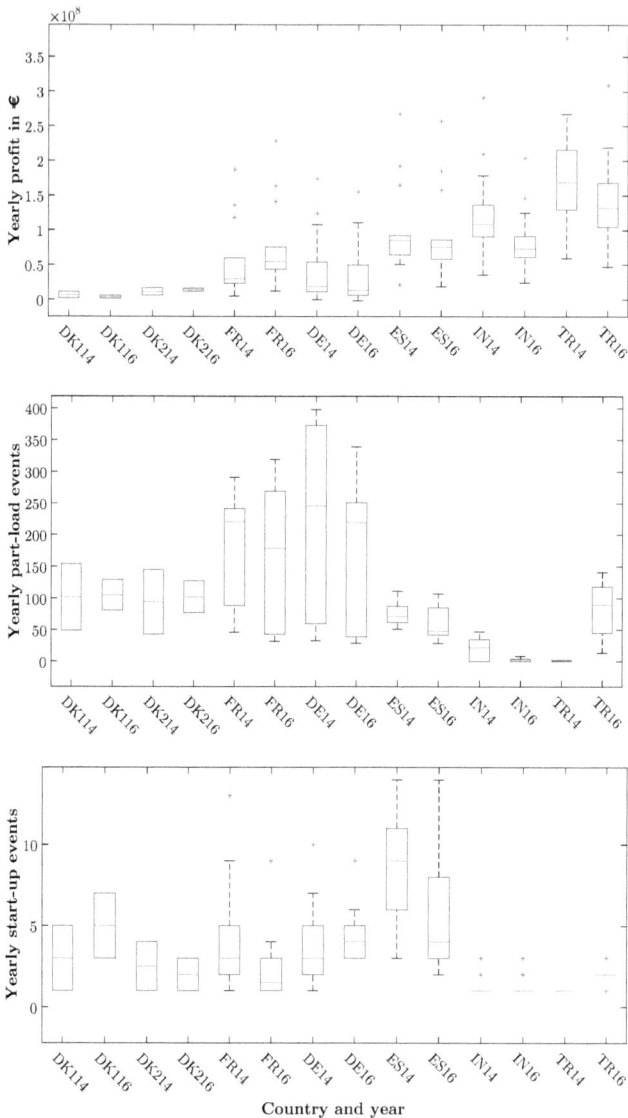

Figure 20 Boxplot analysis of the yearly profit and cycling events for the years 2014 and 2016 in selected market environments. DK: Denmark, FR: France, DE: Germany, ES: Spain, IN: India, TR: Turkey, '14: year 2014. '16: year 2016.

3.1.2 Frequency control capacity reservation

This section analyzes the effects of the capacity reservation for frequency control on the day-ahead dispatch. The case study considers Germany with its secondary control (aFRR) and France with its *réserve rapide* (mFRR). Both require a capacity reservation for the entire peak and/or off-peak period of the week. In Germany, the off-peak period ranges from 20:00 to 8:00 during weekdays and the whole weekend, whereas in France the off-peak period ranges over the weekend only. The case study further considers Spain and its secondary control (aFRR). The case study has been extended by simulating one power plant in the Danish market with its tertiary control market (mFRR). Both the Danish and Spanish simulated markets have an hourly capacity reservation. Denmark West is connected to Germany, whereas Denmark East is connected to Norway and Sweden.

Table 30 summarizes the effect of capacity reservation for frequency control regarding profit and dispatch at a subcritical hard coal-fired power plant (referenced UC-H R in section 2.1) over the whole year 2014. In this case, the yearly profit increase and yearly increase in cycling events due to the capacity reservation are the chosen metrics. The results show that in Germany and Denmark East, the power plant has a similar increase of its day-ahead profit when reserving capacity for the frequency control market. In Spain, it obtains an even greater profit increase by selling capacity in the frequency control market. As combined heat and power plants play an essential role in Denmark, the case study has been extended to include heat extraction. Power plants with combined heat and power have fewer opportunities in the Danish frequency control market. The must-run condition reduces the amount of capacity that can be sold on the frequency control market but also reduces the costs related to spinning control, as the plant has to be online regardless of the day-ahead market outcomes to provide heat. This result shows that, in this context, both effects do not compensate each other. The negative effect of the limited servable capacity is not compensated by the reduced opportunity costs. The impact of the capacity reservation on the power plant dispatch, when compared to the day-ahead dispatch, shows that in Germany alone the number of part-load events reduces when capacity is reserved. This moderate reduction from 212 to 200 yearly part-load events is comparable to the results in Denmark West, where the simulated power plant operates 155 part-load events in the year 2014. When heat introduces a must-run condition, this power plant operates 54 starts a year. In the Spanish context, the power plant cycles a lot more at part-load when capacity is reserved in the frequency market compared to the day-ahead dispatch. The Spanish frequency control prices show a very high variability compared to Denmark so that the power plants can make use of their flexibility in this market. Denmark also has an hourly capacity reservation, but is not symmetric, so that this characteristic might also participate in the drastic increase in part-load cycling in the Spanish context. The profit increase in Germany, when compared to Spain, is rather low. In the French market context, the optimization elicits the decision to not participate in the frequency market at all. The French and the German markets have the common characteristic of a weekly reservation of the capacity. This result leads to the conclusion that this weekly reservation might not be the most suited design for conventional power plants to increase their profit in the frequency control market. The authors of [133] note that this design also represents a drawback for the renewables integration.

Table 30 Day-ahead profit and dispatch modification when capacity is reserved for the frequency control market at a subcritical hard coal-fired power plant (UC-H R) in the year 2014

	Yearly profit increase due to frequency control capacity reservation	Yearly start-up increase due to frequency control capacity reservation	Yearly part-load increase due to frequency control capacity reservation
Germany	4.9%	40%	-5.7%
Spain	18.9%	85.7%	766%
Denmark West	1.4%	0%	0.6%
Denmark West (with CHP)	0.2%	0%	0%
Denmark East	6.4%	25%	6.9%
Denmark East (with CHP)	1.6%	0%	23.3%

The calculations have been extended to the full set of power plant technologies introduced in section 2.1. As visible in Figure 22, frequency control capacity reservation still offers more opportunities in Spain than in Germany, but the profit increase depends on the power plant technologies. When observing the results for Spain, the profit increase in the frequency control market, when compared with the day-ahead profit, is higher with lower minimum load and better ramp rate capabilities. The results, as depicted in Figure 25, indicate that better flexibility parameters lead to better profit increases. For a given minimum load, better ramping capability always leads to better profit increase in the frequency control market. The results for Germany are not as explicit. Some power plants with better ramping and minimum load capabilities do not reach the profit increase reached by less flexible ones. This result indicates that the design of the German secondary frequency control, primarily due to the reservation for a whole week, might not be suited to reward the most flexible power plants. This conclusion is supported by the results in the French market environment. None of the simulated power plants took part in the frequency control capacity reservation in France. The German frequency control, however, allows all simulated power plants to increase their profit. The number of weeks each generation technology reserved capacity in the German frequency control market 2014 and 2016 is depicted in Figure 21. The decreased participation in 2016 compared to 2014 reflects the falling prices for frequency control, often argued as stemming from the increased number of actors offering frequency control. For Germany and Spain, this profit increase is accompanied by an increase in cycling in the power plant dispatch (see Figure 23 and Figure 24). This leads to the conclusion that the frequency control markets are well designed to incent flexible operation with a corresponding reward. In Germany, the profit increase is, however, lower in magnitude in the year 2016 than in 2014, despite growing shares of renewables.

In the Spanish market environment 2016, the hard coal-fired ultra-supercritical power plant USC-H reached less profit when reserving capacity for the frequency control market than when dispatched in the day-ahead market alone. This result shows that despite the improved formulation, the opportunity cost calculation (even with perfect price foresight) does not capture all parameters leading to a comprehensive cost-benefit analysis. The analysis of this power plant's dispatch shows that one of the missing parameters is the increase in cycling induced by this capacity reservation. The dispatch with frequency control has a relative increase of start-ups and part-load events of respectively 767 % and 123 %. These are not included in the opportunity cost calculation for the decision to reserve capacity or not for the frequency control market. An approach combining the opportunity cost calculation and the consecutive simulation of the dispatch might be necessary to reduce the risk related to the decision to reserve capacity or not. However, for most of the cases calculated in this case study, the opportunity cost approach is sufficient to ensure a profit increase.

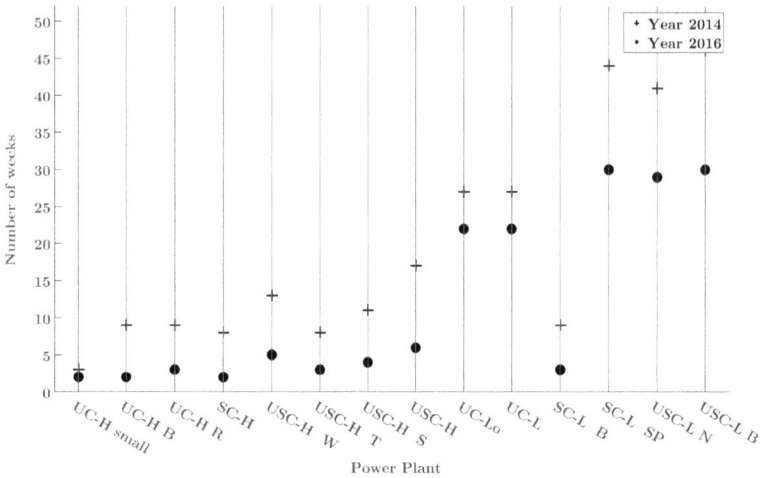

Figure 21 Number of weeks the power plants reserved capacity for the German frequency control market in 2014 and 2016

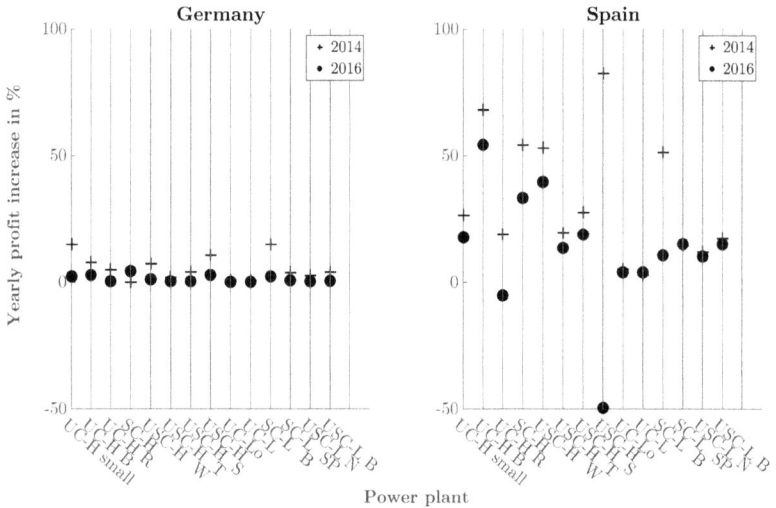

Figure 22 Yearly profit increase due to frequency control at selected power plants dispatched in the German and Spanish market environments 2014 and 2016. The increase is in percent profit increase compared to the day-ahead profit.

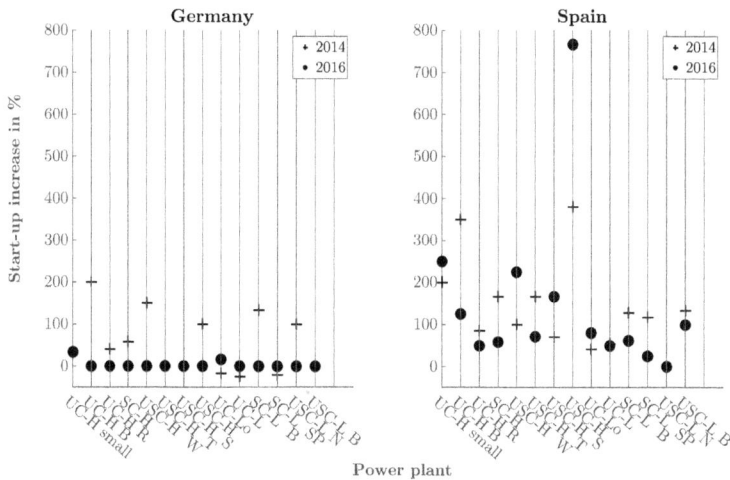

Figure 23 Yearly start-up increase due to frequency control at selected power plants dispatched in the German and Spanish market environments 2014 and 2016. The increase is in percent event number change compared to the day-ahead profit.

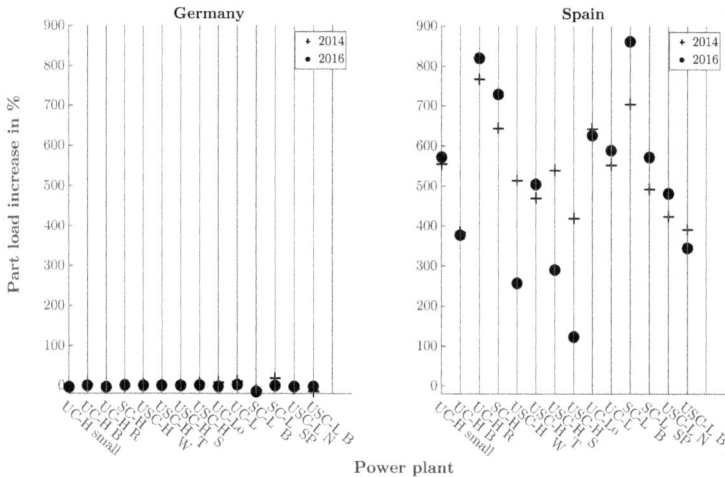

Figure 24 Yearly part-load increase due to frequency control at selected power plants dispatched in the German and Spanish market environments 2014 and 2016. The increase is in percent event number change compared to the day-ahead profit.

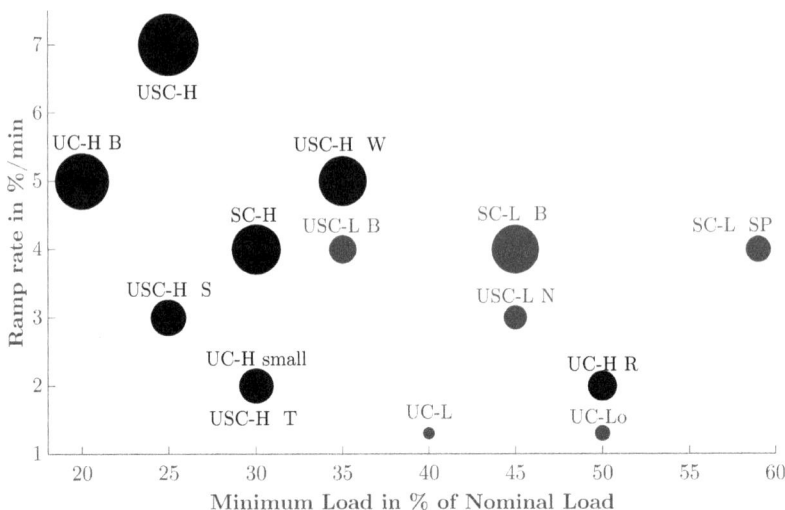

Figure 25 Profit increase reached in the Spanish market environment when offering capacity in the frequency control market in the year 2014. The size of the circles is proportional to the percentage yearly profit increase when compared to the day-ahead profit. For absolute values see Figure 22.

3.1.3 Intraday arbitrage

This section analyzes the effect of intraday arbitrage on the day-ahead dispatch. This case study considers Germany, Spain, Denmark, and France in the year 2014 and Spain, Denmark and Turkey in the year 2016 (due to data availability). Germany has a 15-minute price signal, whereas France, Spain, Denmark, and Turkey have hourly price signals. The calculations are carried out at a subcritical hard coal-fired power plant, referenced UC-H R in section 2.1. The chosen metric is again the yearly profit increase and increase in cycling events due to the arbitrage in the intraday market, compared to the day-ahead dispatch. The results, summarized in Table 31 and Table 32 show that the arbitrage opportunities are very different from a country to another. This power plant does not find arbitrage opportunities in the Spanish market, in 2014 as well as in 2016. In France, the opportunity is moderate and leads to a reduction of start-up events and an increase in part-load events. In Germany and Denmark, the arbitrage opportunities lead to a high increase of the day-ahead profits, but in the case of a must-run condition such as combined heat and power provision, this profit is reduced. In these market environments, must-run conditions reduce arbitrage opportunities. In Denmark, the intraday arbitrage results in start-up and part-load event increases, whereas in Germany the start-up events decrease. When comparing Denmark in the year 2014 and 2016, the two Danish zones show different results. Denmark West is connected to Germany, whereas Denmark East is connected to Norway and Sweden. The year 2016, with increased renewables shares, offers more profit increase opportunities in Denmark West, whereas the contrary occurs in Denmark East. Denmark West also sees a change toward reduced part-load events, as in Germany in 2014. The Denmark East zone, as connected to Sweden and Norway, might be less affected by the renewables volatility, as it is connected to a market with high storage capacity. The Turkish intraday market, put in place in 2015, offers a moderate profit increase, with reduced start-ups and increased part-load.

The calculations have been extended to the power plant technologies introduced in section 2.1, see Figure 26, Figure 27, and Figure 28. These figures illustrate the increase in profit and cycling events reached at each power plant using the intraday market as an arbitrage opportunity. In absolute value, the German 2014 market environment allows for the most significant profit increase via intraday arbitrage. The Spanish 2014 market environment is favorable to the subcritical lignite-fired power plants, which have the lowest ramping capabilities and moderate minimum load capabilities. These results, however, do not hold in the 2016 market environment. The French 2014 market environment allows a broader range of power plants to increase their profit using intraday arbitrage. The lignite-fired power plants are the ones with the lowest profit increase in this context, apart from the lignite-fired plant with better ramping and minimum load capabilities (SC-L B). The Turkish market also allows for a broader range of power plants to increase their profit in the intraday market, but with a lower relative increase compared to France. Lignite-fired power plants are also the ones with lowest profit increase, with again SC-L B as an exception.

Table 31 Yearly profit and cycling increase due to intraday arbitrage in the year 2014 at a
hard coal-fired subcritical power plant (UC-H R). The increase is measured in comparison
to the day-ahead dispatch without intraday arbitrage.

	Profit increase due to intraday arbitrage	Start-up increase due to intraday arbitrage	Part-load increase due to intraday arbitrage
Germany	125.6%	-70%	373.6%
Spain	0%	0%	0%
France	4.4%	-7.7%	7.3%
Denmark West	337.9%	240%	16.9%
Denmark West (with CHP)	17.4%	100%	140.1%
Denmark East	154%	325%	84.1%
Denmark East (with CHP)	12%	100%	190.7%

Table 32 Yearly profit and cycling increase due to the intraday arbitrage in the year 2016 at
a hard coal-fired subcritical power plant (UC-H R). The increase is measured in comparison
to the day-ahead dispatch without intraday arbitrage.

	Profit increase due to intraday arbitrage	Start-up increase due to intraday arbitrage	Part-load increase due to intraday arbitrage
Spain	0%	0%	0%
Turkey	1%	-25%	15.3%
Denmark West	532.5%	157%	-51%
Denmark West (with CHP)	80.2%	200%	-84%
Denmark East	31%	333.3%	78.7%
Denmark East (with CHP)	9.9%	400%	110.4%

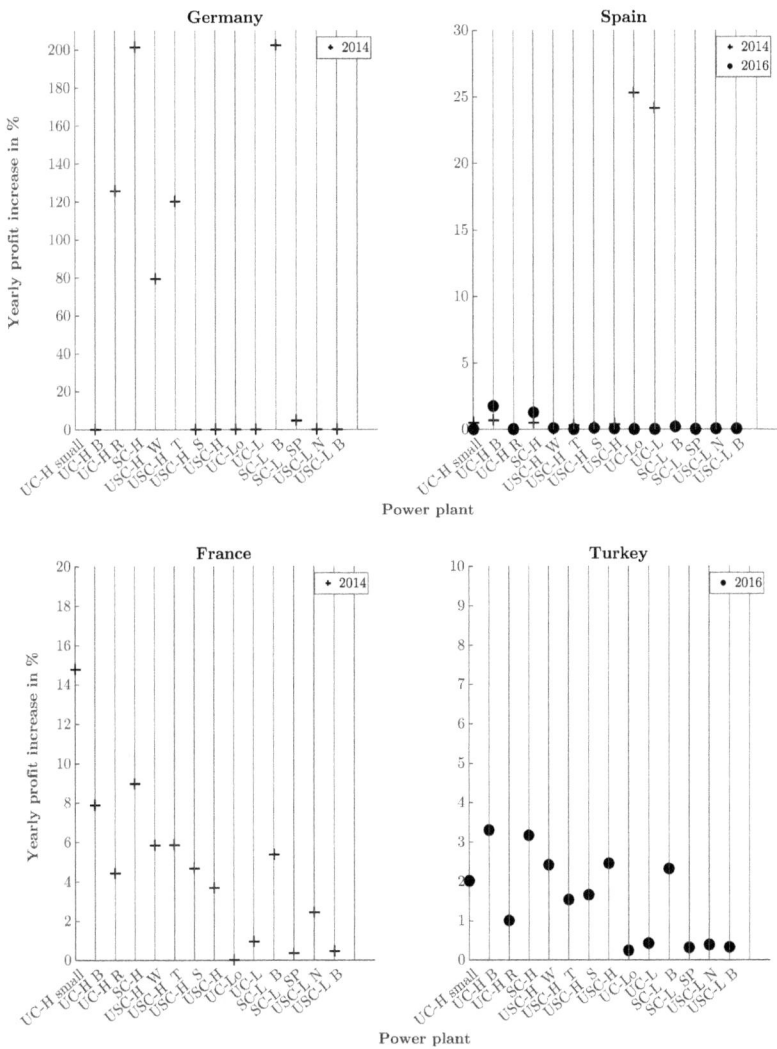

Figure 26 Yearly profit increase due to intraday arbitrage. The increase is measured in comparison to the day-ahead dispatch without intraday arbitrage.

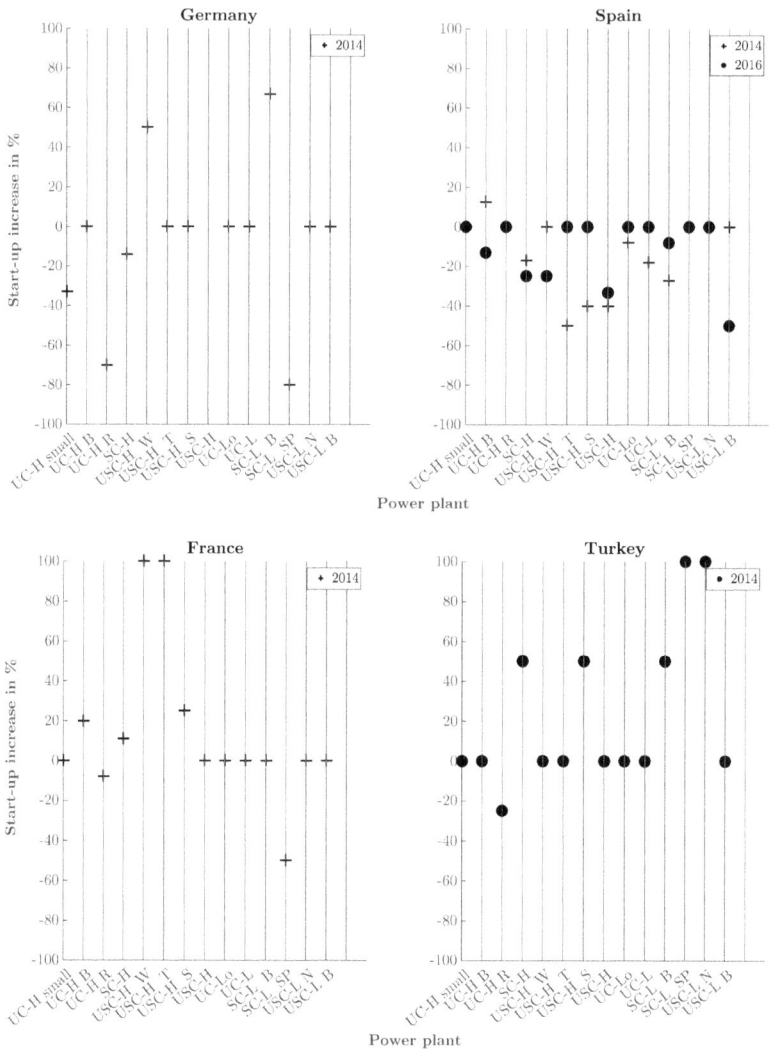

Figure 27 Yearly start-up increase due to intraday arbitrage. The increase is measured in comparison to the day-ahead dispatch without intraday arbitrage.

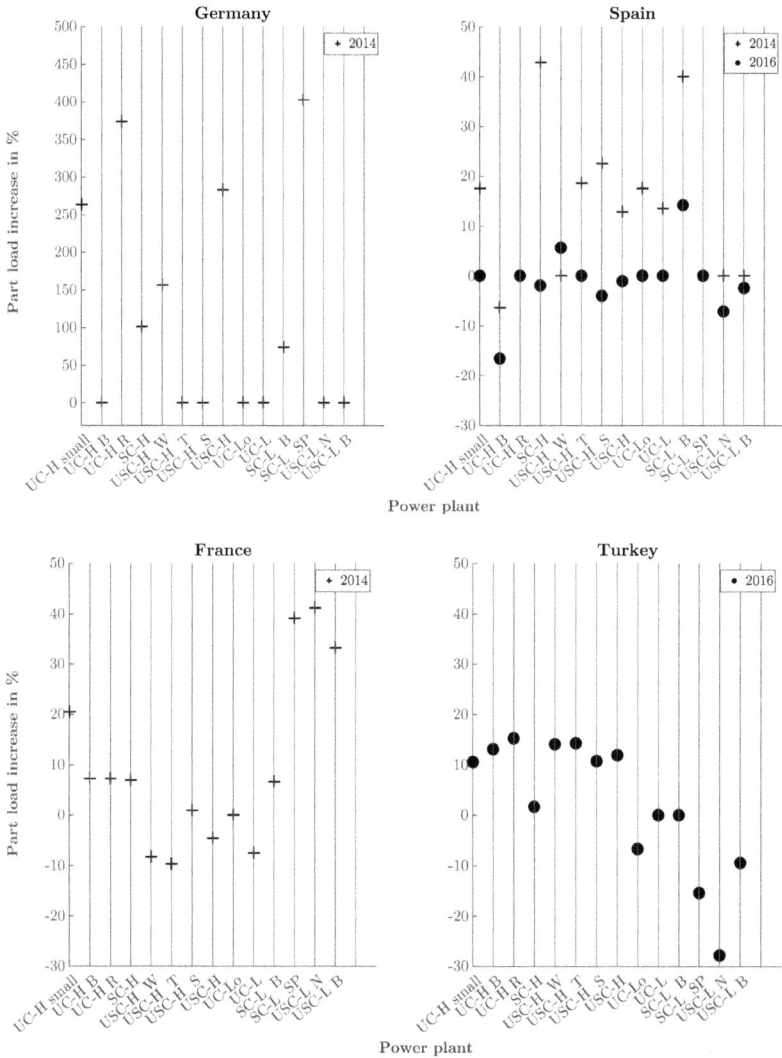

Figure 28 Yearly part-load increase due to intraday arbitrage. The increase is measured in comparison to the day-ahead dispatch without intraday arbitrage.

3.1.4 Combined heat and power

One difficulty related to the simulation of combined heat and power (CHP) plants is the heat price paid to generators, which is not disclosed. For this case study, a heat price of 7.5 and 7.2 €/MWh$_{th}$ is assumed for 2014 and 2016 respectively. The CHP subcritical hard coal-fired power plant (reference UC-H R in section 2.1) is simulated in the German and Danish market environment, and the results are collected in Table 33 to Table 36. The tables show the absolute profit reached and the number of cycles operated by the power plant. The Table 33 and Table 34 show the results with heat extraction, the Table 35 and Table 36 show the results without heat extraction. Some cells are left blank due to data unavailability.

The result shows that the power plant generally made less profit in the year 2016 than in the year 2014 in both markets. The only exceptions occur in the Danish market. The power plant without CHP made more profit in the Danish East market 2016 when compared to 2014. This is also true for Denmark West but only when intraday arbitrage is used. The power plant with CHP makes more profit in the Danish East market 2016 when compared to 2014 at the condition to reserve capacity for the frequency control market. In all other configurations, the yearly profit is decreased when comparing the 2016 market environment to the 2014 one. The CHP power plant in the German context operates fewer start-ups and part-load events in 2016 than in the year 2014, whereas these increase in the Danish market environment. The Danish market allows the CHP power plant to increase its profit using arbitrage opportunities in the intraday market, which is not the case in Germany.

When comparing the profit of the power plant with and without CHP, the results show that the CHP plant earns more money than the non-CHP plant in all market environments with the exception of the German one in 2014, when using intraday arbitrage. Heat, if remunerated at the assumed price levels, thus improves the economic output. In all market environments, the CHP plant operates less cycles than the non-CHP plant, due to the must-run condition. When the power plant already has a must-run condition via heat extraction, the opportunity costs of frequency control provision are much lower. In the German market environment, the CHP power plant thus maximizes its profit with the frequency control market, whereas the non-CHP plant maximizes its profit with intraday arbitrage. In the Danish Western market, the CHP must-run condition does not modify the profit-maximizing market. The same holds for Denmark East, with a slight improvement for the non-CHP plant when using both the intraday and frequency control market. The Danish frequency control market also allows for a profit increase for CHP plants, which is maximized when playing in the three markets. For the non-CHP plant, it seems that playing at the same time in the frequency control market and the intraday market might be sometimes less beneficial than playing in the frequency market alone.

Table 33 Comparison of a subcritical hard-coal-fired (UC-H R) power plant's dispatch with combined heat and power in the German and Danish market environments in 2014

		Germany 2014	Denmark West 2014	Denmark East 2014
Yearly profit in €	Day-ahead	2.01e7	1.09e7	1.65e7
	Day-ahead and frequency control	2.06e7	1.09e7	1.68e7
	Day-ahead and intraday	2.01e7	1.28e7	1.85e7
	Day-ahead, intraday and frequency control	2.06e7	1.28e7	1.89e7
Yearly start-up events	Day-ahead	2	1	1
	Day-ahead and frequency control	3	1	1
	Day-ahead and intraday	2	2	2
	Day-ahead, intraday and frequency control	3	2	6
Yearly part-load events	Day-ahead	97	49	43
	Day-ahead and frequency control	93	49	53
	Day-ahead and intraday	97	118	125
	Day-ahead, intraday and frequency control	93	118	102

Table 34 Comparison of a subcritical hard-coal-fired (UC-H R) power plant's dispatch with combined heat and power in the German and Danish market environments in 2016

		Germany 2016	Denmark West 2016	Denmark East 2016
Yearly profit in €	Day-ahead	1.64e7	6.10e6	1.59e7
	Day-ahead and frequency control	1.69e7	6.27e6	3.56e7
	Day-ahead and intraday	-	1.01e 7	1.74e7
	Day-ahead, intraday and frequency control	-	1.12e7	3.68e7
Yearly start-up events	Day-ahead	1	3	1
	Day-ahead and frequency control	1	3	1
	Day-ahead and intraday	-	9	5
	Day-ahead, intraday and frequency control	-	9	4
Yearly part-load events	Day-ahead	82	81	77
	Day-ahead and frequency control	80	81	68
	Day-ahead and intraday	-	13	162
	Day-ahead, intraday and frequency control	-	13	130

Table 35 Comparison of a subcritical hard-coal-fired (UC-H R) power plant's dispatch without combined heat and power in the German and Danish market environments in 2014

		Germany 2014	Denmark West 2014	Denmark East 2014
Yearly profit in €	Day-ahead	1.27e7	1.93e6	6.13e6
	Day-ahead and frequency control	1.33e7	1.96e6	6.52e6
	Day-ahead and intraday	2.86e07	8.45e06	1.55e07
	Day-ahead, intraday and frequency control	2.84e07	8.40e06	1.36e07
Yearly start-up events	Day-ahead	10	5	4
	Day-ahead and frequency control	14	5	5
	Day-ahead and intraday	3	17	17
	Day-ahead, intraday and frequency control	9	16	27
Yearly part-load events	Day-ahead	212	154	145
	Day-ahead and frequency control	200	155	155
	Day-ahead and intraday	1004	180	267
	Day-ahead, intraday and frequency control	2168	193	257

Table 36 Comparison of a subcritical hard-coal-fired (UC-H R) power plant's dispatch without combined heat and power in the German and Danish market environments in 2016

		Germany 2016	Denmark West 2016	Denmark East 2016
Yearly profit in €	Day-ahead	8.79e6	1.45e6	1.16e7
	Day-ahead and frequency control	8.83e06	1.66e6	3.14e7
	Day-ahead and intraday	-	9.15e06	1.52e07
	Day-ahead, intraday and frequency control	-	9.19e06	3.48e07
Yearly start-up events	Day-ahead	9	7	3
	Day-ahead and frequency control	9	8	3
	Day-ahead and intraday	-	18	13
	Day-ahead, intraday and frequency control	-	20	12
Yearly part-load events	Day-ahead	219	129	127
	Day-ahead and frequency control	212	129	126
	Day-ahead and intraday	-	63	227
	Day-ahead, intraday and frequency control	-	64	218

3.2 Operational flexibility retrofits

This section's focus is on the quantification of the economic value of operational flexibility improvements at thermal power plants. The data collected for the case studies is used to perform exemplary operational flexibility valuations.

3.2.1 Investment decision-making case studies

Two existing German power plants are analyzed using publicly available data and assumptions as detailed in section 2.1 and the Appendix. At each power plant, two retrofit options are tested. The first option (see section 2.2.2) is documented in [167] and [173] for the power plants Rostock (identified as UC-H R) and Schwarze Pumpe (identified as SC-L SP) respectively. The Rostock power plant is retrofitted to achieve an improvement of ramp-rate and minimum load capability. The Schwarze Pumpe power plant is retrofitted to achieve an improvement in the minimum load capability only. The second option tested at these power plants is the indirect firing retrofit presented in section 2.2.1. This retrofit is reducing the plant's minimum load, increasing the ramp-rates and reducing its start-up fuel consumption. For this case study, the selected metric is the net present value (NPV) of yearly cash-flows, used to determine the return on investment (ROI) metric. It is assumed that the investment is considered if it is recovered within three years, as is generally the case in the industry.

For the indirect firing product, the investment is assumed to range approximately between 10 and 60 million euros, see section 2.2.1. These extreme values are used to evaluate the range of return on investment durations shown in Table 38 for the Rostock power plant and Table 39 for the Schwarze Pumpe power plant. A discount rate of 8 % is assumed, and the yearly cash-flows, indicated in the first column, are assumed constant. The results show that if the investment decision had been taken based on 2014 values, it would have been considered only if the plant makes use of the intraday arbitrage opportunities and the frequency control capacity reservation, and with the lowest range of the investment costs. In the year 2016, no configuration leads to considering the investment, even with frequency control capacity reservation. A return on investment of 5 years with a dispatch including frequency control and intraday arbitrage would have been required for the lower bound of the investment to be acceptable in 2016.

As the investment expenditures for the other retrofits are unknown, the same method is used to derive the investment cost limit required for a three-year return on investment. This situation could reflect the decision making of an equipment manufacturer, in the process of investing or not in product development. The upper bound of investment costs is the limit below which planned products would be considered for product development. Above this value, it is unlikely that customers will invest in it, and is thus also not attractive to the equipment manufacturer. The results are summarized in Table 40 and show that the upper bounds of the investment costs all lie below 5 million euros for the selected retrofits. The cells which are left blank are the ones for which the frequency opportunity costs calculation was not sufficient to ensure that the profit is increased (see discussion in section 3.1.1). For these cases, the results do not allow for a conclusion to be formed.

The analysis performed in this section shows that, if three years of return on investment are expected, the current market context will unlikely lead to existing power plant flexibilization investments. In this context, it is more likely that the fleet flexibilization will be limited to the low-hanging fruits of control logic and instrumentation improvements and maybe software retrofits, and thus to a limited flexibilization of the thermal fleet.

The study performed in [84] calculates the yearly profit increase resulting from flexibility improvements at conventional power plants in Germany 2014. The approach chosen there differs from the one used in this work, as the unit commitment problem is solved from a system operator and not power plant operator perspective and as frequency control reserve requirements are defined exogenously. The power system dispatch is optimized to meet the exogenous demand, and the electricity prices are defined endogenously. This study thus provides an interesting comparison basis to the results found in the work at hand. For the year 2014 in Germany, the authors of [84] find that lignite and hard coal-fired power plants have a maximum potential of 6-7 €/(kW*year) profit increase (from low to high flexibility), with standard power plants reaching 4 €/(kW*year). These results include the operation in the frequency control market. For the Rostock power plant (506 MW nominal), this corresponds to 2e6-3.5e6 €/year. For the Schwarze Pumpe power plant (800 MW nominal), this corresponds to 3.2e6-5.6e6 €/year. The value of the considered retrofits at the Rostock (4e6 €/year) and Schwarze Pumpe power plant (6.2e6 €/year) are slightly above the maximum of the value range found in [84], possibly due to, among all other variating parameters, the start-up fuel reduction analyzed in the work at hand not included in [84]. The value of the increased flexibility found by the two methodologies are however in the same order of magnitude. The work at hand offers further insights into the value of operational flexibility increases. These do not only include the plant specific value of flexibility, but also the market dependent value. It is thus interesting to consider the value of the operational flexibility retrofit in the frequency control markets of the other assessed countries. In the Spanish market 2014, this value is of 3.3e7 €/year (6.6e4 €/MW/year) for the Rostock power plant and 3.7e7 €/year (4.6e4 €/MW/year) at the Schwarze pumpe power plant. In the year 2016, this values become 3e7 €/year (5.8e4 €/MW/year) at the Rostock power plant and 2e7 €/year (2.5e4 €/MW/year) at the Schwarze Pumpe plant. In the French market context 2016, these values become 1.4e6 €/year (2.8e3 €/MW/year) at the Rostock power plant and 2e5 €/year (2.5e2 €/MW/year) at the Schwarze Pumpe power plant.

Table 37 Yearly profit increase in the frequency control markets in France, Germany and Spain reached by the indirect firing flexibilisation.

	Rostock-like plant €/MW/year	Schwarze Pumpe-like plant €/MW/year
Germany 2014	7900	7700
Germany 2016	4400	1300
Spain 2014	66000	46000
Spain 2016	58000	25000
France 2016	2800	250

3.2.2 Analysis of the retrofit product's influence on the power plant dispatch

The calculations performed for the investment decision making analysis presented in section 3.2.1 not only result in the economic values presented there, but also in detailed dispatch information. Compared to the single return on investment, net present value or profit increase metrics, the plant's dispatch entails more detail. This information can be used to analyze the behavior of the power plant depending on the market prices, how the profit evolves with the power plant dispatch but also how the dispatch of the power plant is modified when it is dispatched in two or more markets simultaneously. A further use is to compare the operation of two power plants, and especially of a power plant and its retrofitted version. The case study of the Rostock power plant and its retrofit-

ted version in the day-ahead and intraday market is analyzed in
the following. The data in Figure 29 represents the power plant
load (above) and profit (below) from the 2358th to the 2377th hour
of the year 2014. This is the 9th of April 2014 from 05:00 to mid-
night. The baseline power plant data is represented in red, whereas
the retrofit plant is represented in blue. The profit is integrated
every 15 minutes; the time discretization of the German intraday
market. The first figure represents the data when the power plant
is dispatched in the day-ahead market, whereas in the second one,
the power plant also makes use of arbitrage opportunities in the intraday market. The dashed lines represent the evolution of the
day-ahead and intraday spot prices. The load dispatch in the day-
ahead market (first figure) shows how the indirect firing retrofit
modifies the power plant operation during part-load operation. The
deeper minimum load but also the steeper ramp-rates are used to
minimize the losses when bridging the times with lower electricity
prices. The second figure shows how the dispatch is modified when
intraday arbitrage is used. High intraday prices occur in situations
with generation scarcity and incent power plant operation, whereas
low intraday prices incentive power plants to lower their production in times of energy oversupply. The chosen hours span over a
time with high intraday price variations, and the operation profile
shows that the power plant increases and decreases its load as expected. At the beginning of the time horizon, instead of ramping
up, the power plant ramps down as the intraday prices plunge.
When the intraday prices increase, the power plant operates at full
load instead of part-load and sells the additional energy on the intraday market. Instead of just minimizing the losses at part-load,
the power plant thus maximizes its profit using intraday arbitrage.

It is noticeable that at the beginning of the represented time horizon, the retrofitted power plant does not make as much money as
the baseline power plant from buying the surplus energy on the intraday market and parking the power plant at part-load. This is
due to the higher costs incurred for a shallow minimum load operation, which are not compensated by the difference between day-
ahead and intraday prices. The reason for the retrofitted power
plant to operate at its minimum load instead of an intermediary
part-load level is the power plant's state machine definition. As
discussed in section 4.2.1 Part B, there is no economic optimum in
the day-ahead market to operate at part-load. Hence the state-
machine does not define such a behavior. With respect to the state
machine, the work at hand thus provides a lower bound to the
profit increase reached using intraday arbitrage. However, with

regard to the perfect price foresight assumption, the proposed approach provides an upper bound to the profit increase reached using intraday arbitrage.

Table 38 Investment case study for the indirect firing retrofit at the Rostock plant. The interest rate is assumed to be at 8%, and the annual cash flow is considered constant. NPV: Net present value.

	Cash flow: annual profit increase in €	NPV year 2	NPV year 3	NPV year 4	NPV year 5	NPV year 6
Day-ahead 2014	3.16e6	5.64e6	8.14e6	1.05e7	1.26e7	1.46e7
Day-ahead and frequency control 2014	3.97e6	7.08e6	1.02e7	1.31e7	1.59e7	1.84e7
Day-ahead and intraday 2014	4.52e6	8.06e6	1.16e7	1.50e7	1.80e7	2.09e7
Day-ahead, frequency control and intraday 2014	4.53e6	8.08e6	1.17e7	1.50e7	1.81e7	2.09e7
Day-ahead 2016	2.14e6	3.82e6	5.51e6	7.09e6	8.54e6	9.89e6
Day-ahead and frequency control 2016	2.21e6	3.94e6	5.70e6	7.32e6	8.82e6	1.02e7

Table 39 Investment case study for the indirect firing retrofit at the Schwarze Pumpe power plant. The interest rate is assumed to be at 8%, and the annual cash flow is considered constant. NPV: Net present value.

	Cash flow: annual profit increase in €	NPV year 2	NPV year 3	NPV year 4	NPV year 5	NPV year 6
Day-ahead 2014	0.72e6	1.28e6	1.84e6	2.37e6	2.85e6	3.31e6
Day-ahead and frequency control 2014	6.2e6	1.11e7	1.60e7	2.05e7	2.48e7	2.87e7
Day-ahead 2016	0.6e6	1.01e6	1.47e6	1.88e6	2.27e6	2.63e6
Day-ahead and frequency control 2016	1.0e6	1.81e6	2.62e6	3.37e6	4.06e6	4.70e6

Table 40 Investment cost upper bound for a three-year return on investment. The upper bound is calculated as the net present value of the yearly cash flow, which is assumed constant and with an interest rate of 8%.

	Annual profit increase in €	Investment cost upper bound for a three-year return on investment in €
Rostock		
Day-ahead 2014	1.58e6	4.07e6
Day-ahead and frequency control 2014	1.76e6	4.54e6
Day-ahead and intraday 2014	1.58e6	4.07e6
Day-ahead, frequency control and intraday 2014	-	-
Day-ahead 2016	0.94e6	2.42e6
Day-ahead and frequency control 2016	1.76e6	4.54e6
Schwarze Pumpe		
Day-ahead 2014	0.49e6	1.26e6
Day-ahead and frequency control 2014	1.04e6	2.68e6
Day-ahead and intraday 2014	0.25e6	0.64e6
Day-ahead, frequency control and intraday 2014	1.04e6	2.68e6
Day-ahead 2016	0.4e6	1.03e6
Day-ahead and frequency control 2016	-	-

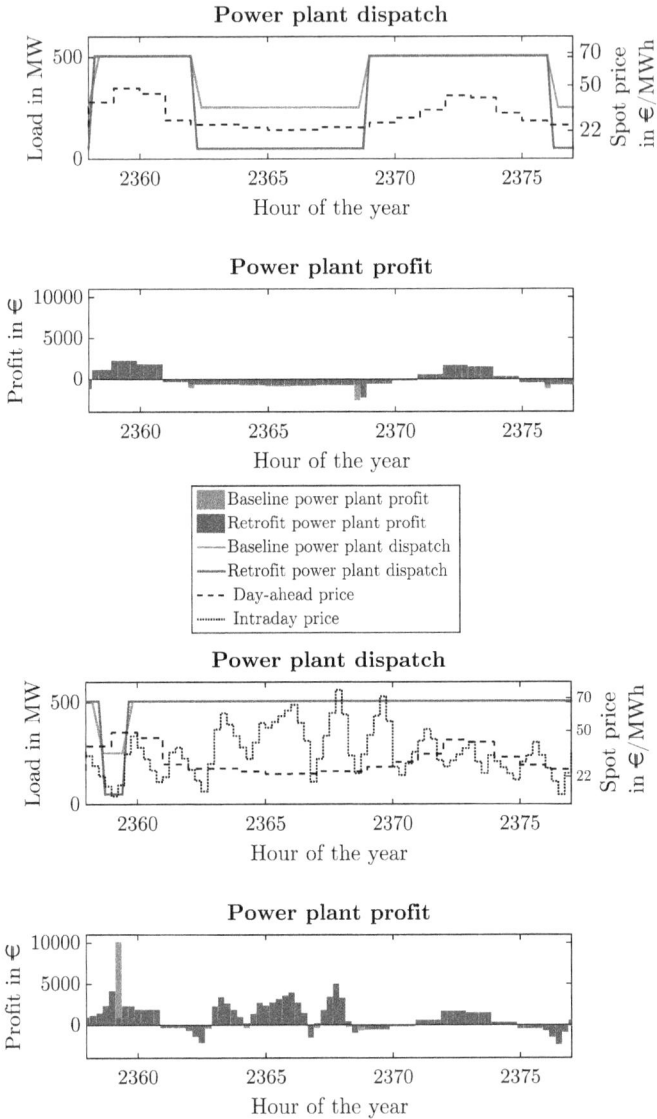

Power plant dispatch

Power plant profit

Legend:
- Baseline power plant profit
- Retrofit power plant profit
- Baseline power plant dispatch
- Retrofit power plant dispatch
- Day-ahead price
- Intraday price

Power plant dispatch

Power plant profit

Figure 29 Power plant dispatch and profit in the day-ahead and intraday markets 2014.
Top: Day-ahead, Bottom: Day-ahead and Intraday.

3.2.3 Analysis of the value of a flexibility product in varying contexts

The value of an operational flexibility option regarding relative profit increase has shown to vary with the power plant technology and market environment, as well as with the dispatch strategy chosen by the operator. This section thus analyses the indirect firing flexibilization option presented in subsection 2.2.1 for the power plants defined in section 2.1 in the market contexts introduced in section 2.3. The dispatch of each power plant and its retrofitted version is optimized for each defined market environment. The results are analyzed using boxplots. The central mark indicates the median, the value separating half of the data sample. The bottom edge of the box indicates the 25th percentile, the value below which 25 % of the data sample is to be found. The top edge of the box indicates the 75th percentile, the value below which 75 % of the data sample is located. The top whisker indicates the maximum, whereas the bottom whisker indicates the minimum of the data sample. Red crosses mark outliers. Outliers are defined as the values which are outside of the range defined by adding (subtracting) one and a half times the box size to the 75th (respectively 25th) percentile.

Figure 30 and Figure 31 compare the value of the retrofit product in terms of yearly profit increase. The obtained yearly profit increases are first analyzed as a function of the market environment and then as a function of the power plant. The increase in profit measures the value of the flexibility improvement in monetary terms. When considering the day-ahead market alone, all market environments, except the German one and to a lesser extent the French one in the year 2014, seem to be almost technology indifferent. This means that the value of the flexibility improvement shows low variations in dependence of which power plant it improves. In the Indian and Turkish market environment, the value of the product is the most technology indifferent. The same conclusion pertains to the analysis including intraday arbitrage. When the capacity reservation for frequency control is included, the Spanish market shows a higher sensitivity to the power plant technology than when the day-ahead market alone is considered. The value range is in this case comparable to the one in Germany. When comparing the value of the flexibility product at the assessed power plants as a function of the market environment (see Figure 31), the results show that for a given plant, the value of the flexibility retrofit regarding relative profit increase is sensitive to the market environment, except for subcritical lignite-fired plants in the day-ahead market.

When comparing the value of the product, which is measured as the profit increase in the year 2014 for all countries, it appears that the product has on average the highest value in Germany and the lowest in Turkey. When selling capacity in the frequency control market, the product has almost same average value in Germany and Spain. When using the frequency control market and the intraday arbitrage opportunities, the product has the highest value in the Spanish market.

When comparing the value of the product regarding profit increase in the year 2016 for all countries, the results show that the flexibility product has the highest value in the German market and the lowest in India, followed by Turkey. When using intraday arbitrage opportunities, the product has similar value in the Spanish and Turkish market. When selling capacity in the frequency control market, the product has the highest value in the Spanish market, followed by the German market environment. The French market environment with frequency control has a low average product value compared to these countries.

The statistical analysis' variability, however, shows that the average tendency might give a truncated picture of the situation. Therefore, the following paragraphs differentiate the average tendency and the power plant specific result.
When comparing the German market context in the year 2014 and 2016, the flexibility product has in average less value in 2016 than in 2014, in the day-ahead market alone as well as in the day-ahead and frequency control market. The situation is, however, different from power plant to power plant. When selling capacity to the frequency control market, the product has more value in 2016 than in 2014 for two power plants (SC-L B and UC-H B). When selling energy on the day-ahead market alone, for a majority (ten in thirteen) of the assessed power plants the product has more value in 2016 than in 2014. This result is in line with the expected behavior with an increased share of renewables in 2016 compared to 2014.
When comparing the Spanish market environment in the years 2014 and 2016, the flexibility product has on average less value in the year 2016 than in the year 2014 when energy is sold on the day-ahead market alone. This is also true when using arbitrage in the intraday market. When selling capacity in the frequency control market, the product has more value in 2016 than in 2014. The

average tendency in the day-ahead market holds for all plants, but three hard coal-fired power plants (UC-H small, UC-H B, SC-H). The average tendency in the frequency control market holds for only three power plants (UC-H R, USC-H, SC-L B). These results are in line with the expected behavior with the stable, slightly decreased share of renewables in 2016 compared to 2014.

When comparing the French market context in the year 2014 and 2016, the flexibility product has, on average, less value in 2016 than in 2014. This average tendency applies to all plants but the USC-L B plant.

When comparing the Turkish market context in the year 2014 and 2016, the flexibility product has on average more value in the year 2016 than in the year 2014. This average tendency applies to all studied power plants. The Turkish intraday market was introduced in 2015 so that no comparison between 2014 and 2016 can be made in this market. However, the calculation shows that, for the year 2016, the flexibility product has more value if intraday arbitrage is used. These results are in line with the expected behavior with the increased shares of renewables.

When comparing the Indian market context in the year 2014 and 2016, the flexibility product has on average less value in the year 2016 than in the year 2014. This average tendency applies to half of the simulated power plants. The share of renewables was almost the same in 2016 as in 2014.

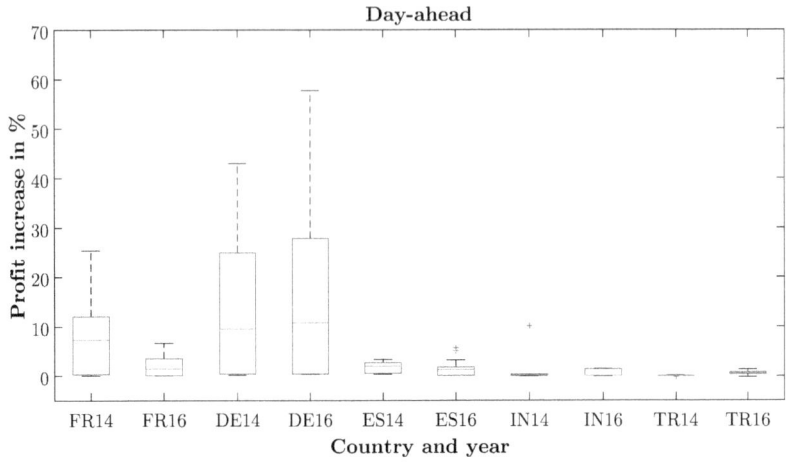

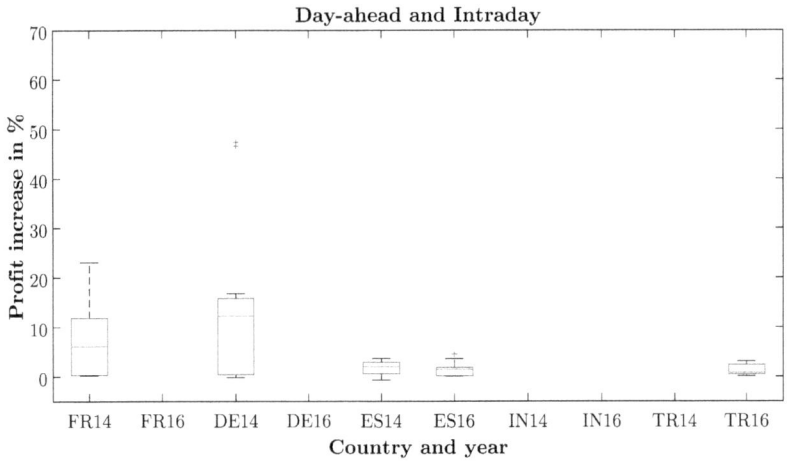

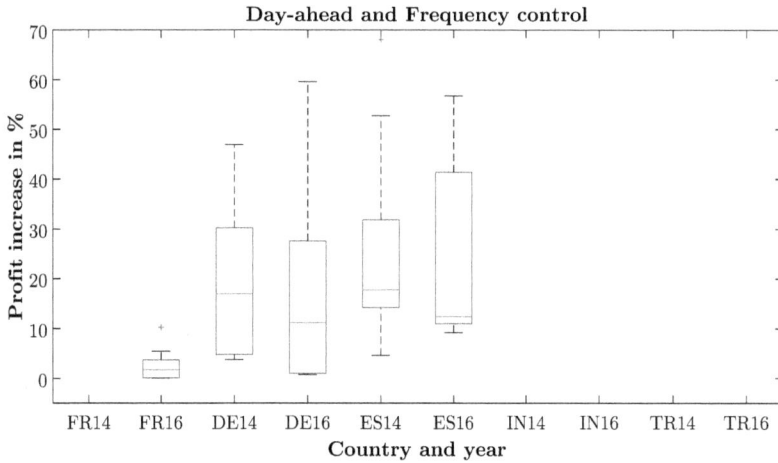

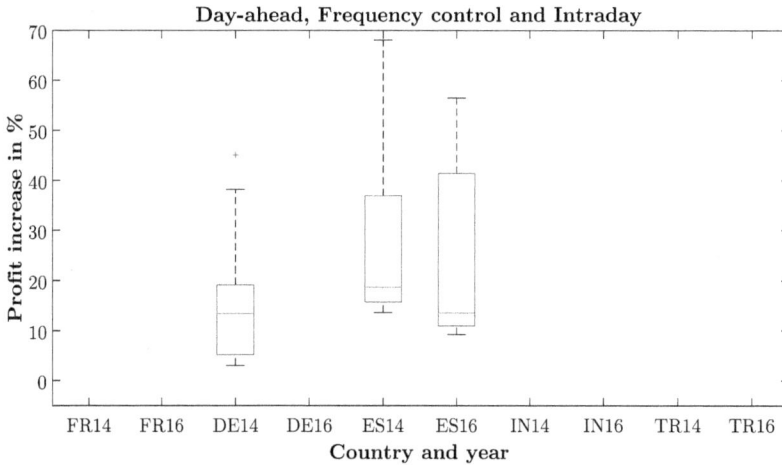

Figure 30 Relative profit increase reached with an indirect firing retrofit as a function of the power plant for different market environments. Boxplots indicate the median with the central mark, and the bottom and top edges of the box indicate the 25th and 75th percentiles, respectively. The whiskers extend to the most extreme data points, and the outliers are plotted individually using the '+' symbol. Some outliers have been excluded from the graph for readability. FR: France, DE: Germany, ES: Spain, IN: India, TR: Turkey.

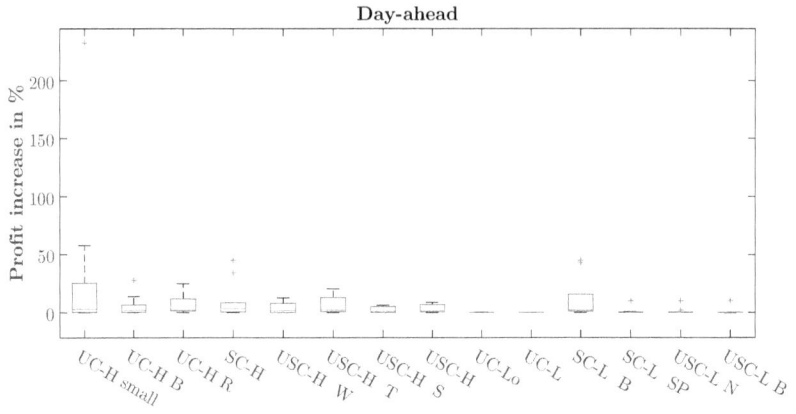

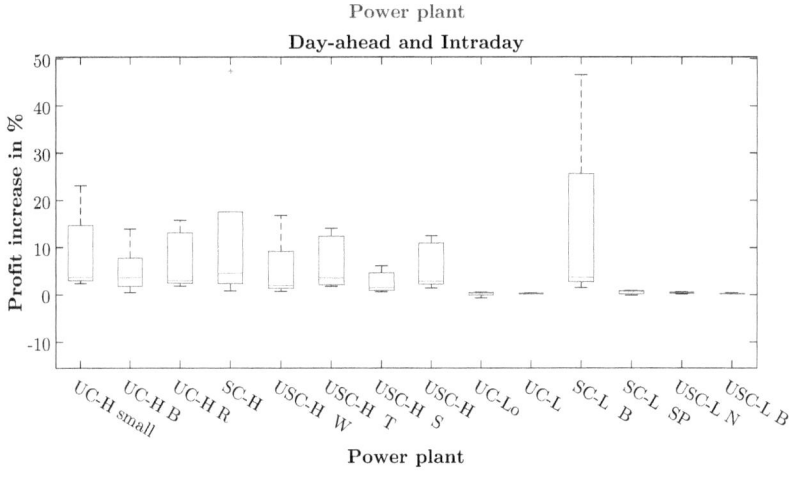

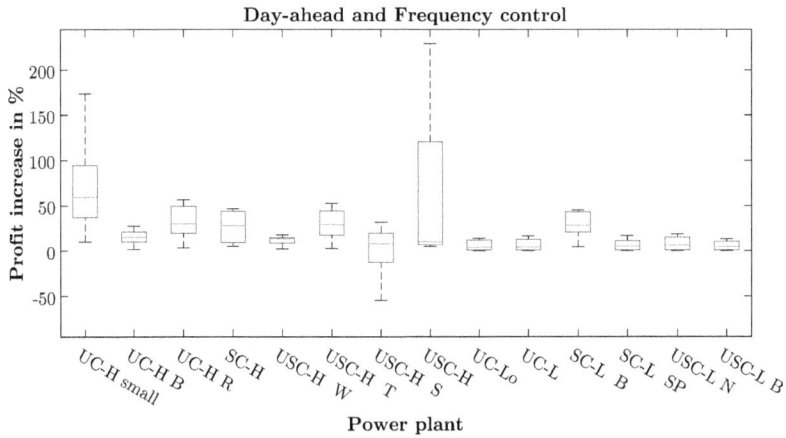

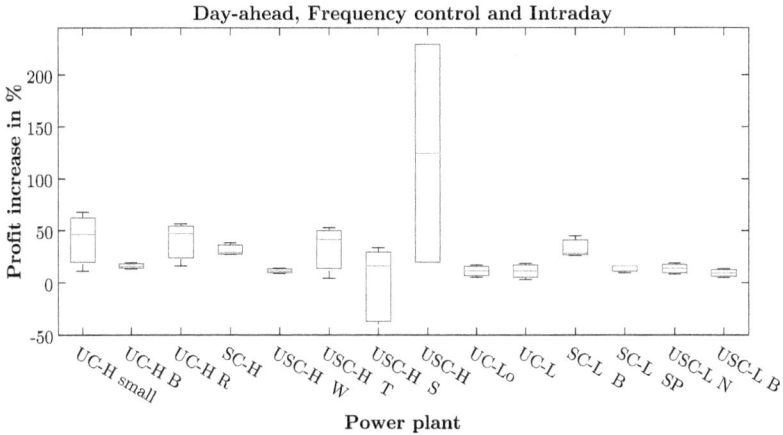

Figure 31 Relative profit increase reached with an indirect firing retrofit as a function of the market environment for different power plants. Boxplots indicate the median with the central mark, and the bottom and top edges of the box indicate the 25th and 75th percentiles, respectively. The whiskers extend to the most extreme data points, and the outliers are plotted individually using the '+' symbol. Some outliers have been excluded from the graph for readability. FR: France, DE: Germany, ES: Spain, IN: India, TR: Turkey.

3.2.4 Analysis of the value of generic flexibility improvements in the German context

As the indirect firing case has shown, a given flexibility improvement is likely to impact more than one operational flexibility attribute (minimum complaint load, ramp-rate, etc.) at a time. Therefore, this section analyses the single and combined impact of generic flexibility improvements on the flexibility attributes minimum complaint load, ramp-rates, and start-up ignition fuel consumption. The ramp-rate increase from 3 %/min to 6 %/min is modeled with an increase in maintenance cost by a factor 1.2 as in [124]. It might be realized by retrofitting thick-walled components at the power plant. The minimum complaint load reduction down to 10 % is technically feasible via indirect firing or one mill operation. This last option might seem unlikely to be used due to the missing redundancies in case of mill shutdown. However, [51] reports that the German power plants Heilbronn 7 and Bexbach use one-mill operation commercially since 2011. Regarding the third option, one half of the baseline power plant's ignition fuel consumption is provided by coal at the retrofitted power plant, thus reducing the start-up costs. The analysis is made at a hard coal-fired supercritical power plant. The power plant parameter values are the same as for the USC-H plant, but with better start-up characteristics (lower ignition fuel consumption and shorter start-up times) and lower ramp-rates. The market environment is the German day-ahead market in the years 2014 and 2016. The results of this analysis regarding relative yearly profit increase and cycling are to be found in Table 41.

Despite the increase in variable renewable-based electricity generation in the year 2016 compared to 2014, the profit increase of all assessed improvements is lower in 2016 than in 2014. The value of the flexibility improvements is higher when capacity is reserved for the frequency control market than when energy is sold on the day-ahead market only. In the day-ahead market, the drastic reduction of the minimum load to 10 % of the nominal load is shown to have the highest relative profit increase, especially when combined with a ramp-rate increase. When capacity is additionally reserved for the frequency control market, the ramp-rate product leads to the highest relative profit increase.

Table 41 Value of generic flexibility improvements in the German day-ahead market 2014 and 2016 measured as the yearly profit increase

	Yearly profit increase in % in the day-ahead market 2014	Yearly cycling events increase in % in the day-ahead market 2014	Yearly profit increase in % in the day-ahead market 2016	Yearly cycling events increase in % in the day-ahead market 2016
Minimum load improvement from 25% nominal load to 10%	1.17%	-66% start-ups +14% part-load	0.60%	-42% start-ups +13% part-load
Ramp rate improvement from 3%/min to 6%/min	0.18%	+33% start-ups -2% part-load	0%	None
Minimum load and ramp-rate improvement	1.30%	-66% start-ups +6% part-load	0.70%	-33% start-ups +8% part-load
Start-up ignition fuel reduction by one half covered by main fuel	0.23%	+55% start-ups -10% part-load	0.06%	+8% start-ups -1% part-load

Table 42 Value of generic flexibility improvements in the German market 2014 and 2016 with capacity reservation for the frequency control market measured as the yearly profit increase

	Yearly profit increase in % in the frequency control market 2014	Yearly cycling events increase in % in the frequency control market 2014	Yearly profit increase in % in the frequency control market 2016	Yearly cycling events increase in % in the frequency control market 2016
Minimum load improvement from 25% nominal load to 10%	1.37%	-50% start-ups +16% part-load	0.60%	-36% start-ups +11% part-load
Ramp rate improvement from 3%/min to 6%/min	13.51%	-17% start-ups +1.5% part-load	2.17%	-9% start-ups +4% part-load
Minimum load and ramp-rate improvement	11.51%	-42% start-ups +6% part-load	1.70%	-36% start-ups +6% part-load
Start-up ignition fuel reduction by one half covered by main fuel	0.22%	+8% start-ups -8% part-load	0.06%	+9% start-ups -1% part-load

3.3 Answering the research questions

This section uses the performed case studies as well as the conceptual work to answer the research questions defined in section 1.4, Part A. The question of whether conventional power plants are technically able to provide operational flexibility but, even more, if the market environment incents the required flexible dispatch has been raised in section 1.4, Part A. Many studies assess the need for flexibility (see review in section 2.1.1, Part A), but how can it be sure that the incentives are set up to encourage the investment in the required flexibility in a liberalized context? Are the markets well designed for flexibility incentives? In the current European context, there will be few or no investments in new thermal power plants, but retrofits for compliance to regulatory constraints or to sell (and make money with) flexibility might come into question: so what is a good flexibility improvement from a commercial point of view? What is the value of conventional power plants' operational flexibility? From a methodological point of view, how can the value of operational flexibility be quantified? How do the chosen power plant model and problem formulation influence the value of operational flexibility?

3.3.1 How to value operational flexibility?

The review in section 2.1, Part A has shown that flexibility assessments consist of two tasks: quantify the need for flexibility and assess the available flexibility to state about its adequacy. This work has developed methods to perform the second task for the case of conventional power plants in a liberalized market environment. As pointed out in [23], the valuation of flexibility by means of a single metric combining different parameters introduces a loss of generality. As such parameters are often related, such a metric would necessarily conceal their interactions [23]. Unit commitment and economic dispatch has shown to be a suited methodology to quantify power plant flexibility, as it dynamically accounts for the power plant's technical and economic constraints. The resulting power plant dispatch provides accurate and detailed information, which can be used to define a suited metric. This work has shown that depending on the point of view (social welfare maximization or profit maximization), the suited problem formulation might differ. In both cases, the problem formulation should include a power plant description accounting for the technical and economic parameters affected by operational flexibility. The time representation has also been shown to influence the quantification of operational flexibility. The case studies presented in this work have made use of the relative profit increase, the absolute profit increase, especially for investment decisions, as well as the cycling behavior of the plant to characterize and quantify the value of operational flexibility.

3.3.2 How do the chosen power plant model and problem formulation influence the operational flexibility value?

The chosen time representation, as shown when analyzing the quality of the event-based approach, influences the operational flexibility assessment. The approximation made when using a time discretization can be reduced via a finer time discretization or the introduction of a continuous representation of time, which requires the concept of events. The quantification of operational flexibility improvements in the intraday, frequency control and combinations of these markets demonstrated the importance of accounting for these markets for a comprehensive analysis. A power plant description with enough detail for operational flexibility valuations but simple enough to be used in the operations research's usual problem formulations has been proposed. The value of flexibility in terms of profit has shown high dependency on the efficiency at part-load. Therefore, when including the frequency control market in the unit commitment and economic dispatch problem formulation, the state-of-the-art opportunity cost calculation has been improved to account for the efficiency degradation. When comparing the insights obtained by the work at hand with the insights offered by e.g. [84], which uses the perspective of the system operator instead of the plant operator, two conclusions might be drawn. The first conclusion is that the value (in Germany, year 2014) of the analyzed flexibility retrofits in the work at hand and in [84] are in the same order of magnitude. The flexibilization value range from a plant operator's perspective (between 3000 and 9000€/MW/year) and from a system operator's perspective (between 4000 and 7000€/MW/year) are consistent. The second conclusion to be drawn is that the work at hands offers more detailed and power plant specific information, which is necessary for investment decision-making from an operator's perspective. Reaching the same level of detail using the method applied in [84] would require extensive data and introduce computational burden.

3.3.3 What is the value of operational flexibility at conventional power plants?

As the case studies show, the value of operational flexibility, and of operational flexibility improvements in terms of relative profit increase and cycling behavior varies from power plant to power plant and from one market environment to the next. Even for power plants with similar flexibility characteristics in a given market environment, the value of an operational flexibility improvement might differ due to further power plant characteristics. The generic flexibility product analysis has shown that the best option in the German day-ahead market is a combination of minimum load reduction and increased ramp-rates, whereas with frequency control the best option improves ramp-rates alone. The investment case studies show that the absolute profit increase reached via the tested retrofits are seldom compatible with the industry standard three-year return on investment requirements. In this context, it is likely that among all possible improvements only the low-hanging fruits, like instrumentation and control, will come into question. When capacity is reserved for the frequency control market, the flexibility that can be offered on the day-ahead (and intraday) markets is reduced, at least in amplitude. The results, however, show that this market increases the value of operational flexibility, when compared to dispatches in the day-ahead market alone. Other streams of revenues, such as intraday arbitrage or capacity reservation for frequency control, help to increase the value of flexibility improvements measured in terms of relative profit increase. These markets should be part of operational flexibility assessments. This work proposes a way to account for these markets from a price-taker perspective, and the results show that, despite improved formulations, opportunity cost approaches for frequency control capacity reservation might not suffice to reach a profit increase. A simulation of the power plant dispatch as performed in the work at hand reduces the risk faced by the operator as it accounts for the cycling of the power plant; a parameter left out by the opportunity costs calculations.

3.3.4 Are the markets well designed for flexibility incentives?

The day-ahead market, when comparing the years 2014 and 2016, incents the power plants to operate more flexibly with increasing renewables share. It does not, however, always compensate for the profit difference resulting from this system compliant dispatch. The frequency control and intraday markets have thus been analyzed. Both markets allow increasing the yearly profit and the value of operational flexibility improvements. The analysis of the indirect firing operational flexibility improvement in the day-ahead markets has shown that the German market environment is the one with the most technologically differentiated incentives regarding relative profit increase. When extending the analysis to frequency control, the Spanish market environment sends more technology differentiated signals than in the day-ahead market alone. In this market context, the power plants with the lowest minimum load and highest ramp-rates reach the highest profit increase, a conclusion that couldn't be raised for Germany, probably due to the weekly capacity reservation. The design of the German secondary frequency control and French tertiary control market with a weekly capacity reservation does not reward the most flexible plants. The intraday markets, aimed at correcting the forecast errors, send the right incentive for conventional power plants to adapt their dispatch to system requirements and represent a source of potential additional revenues. Despite the relatively low intraday markets liquidity, the growing participation to intraday markets in Europe seems to indicate that this potential has been recognized.

3.3.5 What about the environmental impact of flexibility?

Social acceptance is not addressed in this work, but as the nuclear power exit has shown in Germany, it is a decisive aspect of energy problematics. The security aspect of the other conventional power plants is less discussed than in the case of nuclear power. The main drawback and concern associated with coal and gas-fired plants is the environmental threat of emitted greenhouse gas pollutants. While coal-fired power plants are equipped with SOx and NOx filter systems, it is not the case for CO_2. The capture of CO_2 is technically feasible, but the question of the re-use or storage still divides, and most research projects have been stopped in Europe. As the efficiency of power production varies with load, so do the specific CO_2 emissions. The review in [15] suggests that at part-load, coal-fired plants have lower specific SOx, NOx, and CO emissions but higher specific CO_2 emissions than gas-fired power plants. The review in [14] states that emission control of SOx and NOx is technically feasible, even by flexible plant operation, and these pollutants are therefore not included in the model proposed in this work. To account for the CO_2 emissions even by flexible plant operation, the load-dependent efficiency and the fuel specific emission costs and amounts are made use of. As suggested in, e.g. [200], despite the increased specific CO_2 emissions at part-load, flexible plant operation might not impact the absolute emissions, as part-load operation also results in lower total electricity production. The analysis performed in [200] leads to the conclusion that the higher specific CO_2 emissions due to flexible operation do not offset the emission reductions reached via increased wind-based electricity generation. [51] finds lower emissions when operating a retrofitted power plant, compared to its baseline, less flexible version. This conclusion is supported by the calculations performed in this work's case studies. The market environment leading to the steepest increase in start-ups and part-load events when comparing 2016 to 2014 has been selected and analyzed in terms of CO_2 emissions. The dispatch of the hard coal-fired power plants in this environment shows that, despite increased cycling, the absolute CO_2 emissions are stable or even decreased. The specific emission increase at part-load is offset by the reduced absolute electricity production. This result depends on complex interdependencies, but in the case analyzed in this work, the flexible operation leads to more ignition fuel consumption, which has lower specific emissions than coal (more start-ups) and less absolute main fuel consumption (more part-load), despite the degraded efficiency.

4 Conclusion and outlook

A methodological framework to assess the value of a conventional power plant's operational flexibility to support the integration of increasing shares of renewables in the electricity production mix has been developed. As pointed out in [23], the valuation of flexibility by means of a single metric requires assumptions on how to represent crucial parameters like time and costs, thus introducing a loss of generality. As aforementioned parameters are often related, such a metric would necessarily conceal their interactions [23]. Therefore, the proposed work leaves the choice of the metric open. The case studies have shown how to adapt the metric choice to the assessment needs. When applying the developed tools to an investment decision, the net present value and return on investment metrics have been made use of. When assessing the incentives sent from markets with increased shares of renewables regarding flexible operation, the yearly increase in part-load events has been calculated. A specification of the solution methodology to provide accurate quantifications of these metrics has been shown to require high time discretization and technical resolution and long time horizons. Further a requirement is to account for dynamic interactions, such as the power plant's operation costs and market prices or the different time discretization and horizons of intraday and frequency control markets. To fulfill this specification, the work at hand makes use of the unit commitment and economic dispatch problem to quantify the value of operational flexibility. Due to the research questions, the single unit self-scheduling problem of a merchant power plant has been selected and formulated from a price-taker perspective. This formulation includes a power plant model suited for operational flexibility assessments. The quantification of its influence on the power plant dispatch has shown the importance of efficiency degradation at part-load. The problem formulation further makes use of the concepts of events and finite state machines, taken from the Discrete Event Systems theory. This alternative approach to the uniform discretization of time overcomes the finer time discretization required for better solution quality. The problem formulation is extended to account for frequency control capacity reservation, with an improved opportunity cost calculation, and intraday arbitrage. A flexibility improvement's value quantification has shown that these markets influence the power plant's flexible dispatch and monetary incentives to do so. Flexibility studies should, therefore, account for these markets.

The case studies performed in the third part provide detailed insights into the power plant's flexible dispatch and the incentives sent by the various market environments. The observation of the day-ahead markets has shown that these markets send incentives to operate in a more flexible way when the renewables share increases but do not compensate for this system-friendly dispatch. This work does not answer the question of whether this compensation should be put in place, see, e.g., [62]. It has been observed that the frequency control and intraday markets allow for a profit increase when compared to the day-ahead market. The design of the frequency control markets with a weekly reservation does not incent the most flexible power plants. Shorter time horizons should be put in place when pursuing this aim. This change in design would also ease the participation of non-dispatchable renewable plants [133]. The frequency control and intraday markets also increase the value of operational flexibility improvements. The value of such improvements has been studied, with the conclusion that it is market and technology dependent. At some plants, the flexibility improvements have less value despite more volatility in the market, whereas the contrary is observed at other power plants. Furthermore, within a given market environment, the frequency control markets the plant is selling its capacity to and the arbitrage opportunities offered by the intraday market influence which flexibility improvement is best suited. For the German day-ahead market, a reduction of minimum load and increase of ramp-rates is best suited, whereas better ramp-rates only are best suited if capacity is reserved for frequency control. The case study results demonstrate the need for case-specific detailed assessments like the ones enabled by the developed concepts.

The value of operational flexibility retrofits found via the developed approach has also been compared to the values delivered by an alternative approach. The said approach to be found in [84] solves the problem from a system operator's point of view instead of a power plant operator's point of view. The data intense model does not allow as detailed power plant representations as the one used in this work. Despite the different approaches, power plant descriptions and assessed flexibility retrofits, the found values correspond with the range indicated in [84]. Despite values being power plant and market specific, the work at hand provides insights in the value range of operational flexibility retrofits in various countries. At the hard coal-fired power plant similar to the German Rostock power plant, the indirect-firing-like retrofit is worth between 3e3 and 7e4 €/MW/year in the frequency control market, depending on the assessed country. At the lignite-fired power plant similar to the German Schwarze Pumpe power plant, the indirect-firing-like retrofit is worth between 3e2 and 5e4 €/MW/year in the frequency control market, depending on the assessed country.

The case studies presented in the third part are not a comprehensive representation of the applicability of the concepts developed in this work.
Flexibility assessments have shown to require the unit commitment problem extension to frequency control markets. The opportunity cost calculation for frequency control capacity reservation, allowing for a better representation of the losses at minimum load, might be used in other unit commitment problems rather than the one studied in this work. The resulting capacity reservation decisions might be used as an exogenous parameter for the standard mixed integer programming problem, which constraints would need to be adapted to account for the reduced operating range. Formulations for the energy provision with known demand could be taken from, e.g. [140].
It has further been observed that flexibility assessments require power plant models accounting for, at least, efficiency degradation at part-load, non-constant start-up costs and fuel specific CO_2 emission factors. The parametric power plant model has been formulated in a mixed-integer programming problem when validating the event-based approach. It could be integrated into most energy system models, and also be used for studies considering the whole energy system with aggregated technologies from a welfare point of view (whereas this work has a profit maximizing approach considering one plant in detail).

Most of the possible improvements and further research work have been mentioned when reviewing the validity of this work's assumptions (see section 4, Part B). The most significant of these outlooks is to include uncertainty and thus stochasticity within the model. This additional feature would allow risk-averse users to test a broader range of possible outcomes. A straightforward way to implement this is to apply the methods proposed in the work at hand and run the simulations for different inputs with given probability distributions. A more sophisticated alternative is to use the proposed power plant and market models in a stochastic formulation of the dispatch problem. When accounting for uncertainty, it might become worth it to adapt the state machine to allow for part-load operation above the minimum load level when using the intraday market as an arbitrage opportunity (see discussion in section 3.2.2

Part B). For the intraday dispatch, it would also be interesting to measure how the uncertainty and risk evolve the closer the forecast gets to the market lead time. A further improvement proposition, if power plant operators keep large-scale fleets in a system with higher shares of renewables, is to extend the performed case studies with power plant operator arbitrage in its fleet. The results might thus provide a measure of the value of operational flexibility in dependence of the fleet size.

The methodology in the work at hand makes use of the disparity of renewables shares across Europe to base the case studies on historical observations, instead of relying on assumptions and scenarios. In case this alternative methodology was to be chosen, it would be interesting to use the output price time series from fundamental market models as input prices for the models developed in the work at hand. A further step would be to compare the price time series resulting from fundamental models with the ones resulting from agent-based simulations, regression analysis and artificial intelligence models.

The concepts and case studies depicted in this work are intended for the evaluation of the conventional power plant's contribution to the energy transition in terms of operational flexibility. This evaluation is based on the current design of the European markets. The power plant operation strategies assumed in this work are likely to apply in liberalized markets based on a mix of volatile and dispatchable plants. It is, however, questionable, whether these design and operation strategies will apply to a system with a dominant share of renewables and only a few conventional power plants. It might be that in such a system, the few remaining conventional power plants operate only in the case of very critical scarcity situations, providing a form of cold reserve. Conversely, it might be that these few plants are operated continuously, to avoid cycling costs and provide a reliable basis for spinning control. Whatever this system might look like, new concepts for rewarding these services will be required. As the system shifts from a fuel-intensive toward a capital-intensive paradigm [201], it is unlikely that a marginal-cost based price settlement will set appropriate operation and investment signals. In this regard, the authors of [201–203] deliver insights into market designs for short-term supply and long-term capacity adequacy in a fully renewable system. [202] identifies three adaptation options. The first two concepts involve using the concept of virtual power plants to make the new system as was used in the historic system or to adapt the current market design. This adaptation is also addressed in e.g. [201]. The scope of these adaptations concerns itself with the switch from zonal to nodal pricing, the extension from energy-only markets to energy and capacity markets, the introduction of complex bids, as well as an improved time granularity in market design. [203] defines which of these options fit best for future transformative society scenarios. These scenarios reflect deep societal changes in terms of ecological awareness and business structures. [201] suggests that these future energy systems will vary in the degree of government control, and [202] goes further in its third adaptation path by suggesting that state monopolies might be reintroduced. Alternatively, the author of [202] proposes to introduce a pool market as in the North-American markets. This market is characterized by nodal pricing and simultaneous clearing of energy and frequency control. Long-term tariffs for all generation technologies or generation technology-specific auctions with long-term contracts are further suggested. As this overview illustrates, various design options are available and might inspire further design improvements. In this regard, energy systems with already high shares of renewables provide a living laboratory for other countries to go ahead with their energy transitions.

Appendix:
Power plant description

Table 43 Power plant description - General data

Plant ID	Type cluster	Main fuel type	Ignition fuel type	Nominal Load MW	Nominal Efficiency	Minimum Load %	Minimum load efficiency
UC-H small	Subcritical	Hard coal	Natural gas	250MW	36%[163]	30%[164]	32%[165]
UC-H B	Subcritical	Hard coal[166]	Fuel oil	660MW[166]	39.4%[166]	20%[166]	34%[166]
UC-H R	Subcritical	Hard coal[[167]	Fuel oil	506MW[167]	43.2%[167]	50%[167]	38.8%[165] 40.6%[167]
SC-H	Supercritical	Hard coal	Natural gas	800MW	40%[163]	30%[164]	38%[169]
USC-H W	Ultra-supercritical	Hard coal[166]	Natural gas	725MW	46%[166]	35%[166]	43.5%[169]
USC-H T	Ultra-supercritical	Hard coal[166]	fuel oil	660MW	43.7%[166]	30%	41.5%[169]
USC-H S	Ultra-supercritical	Hard coal	Natural gas	500MW	46.6%[166]	25%	41.4%[166]
USC-H	Ultra-supercritical	Hard coal	fuel oil	400MW	46%[163]	25%[164]	40%[169]
UC-Lo	Subcritical	Lignite	Natural gas	400MW	31%[163]	50%[164]	29%[167]
UC-L	Subcritical	Lignite	Natural gas	400MW	31%[163]	40%[164]	29%[167]
SC-L B	Supercritical	Lignite[166]	Fuel oil	858[166]	41%[166]	45%[166]	39.4%[169]
SC-L SP	Supercritical	Lignite	Fuel oil	800MW[173]	41.2%[173]	59% [173]	40.4%[169]
USC-L N	Ultra-supercritical	Lignite	Fuel oil	1100MW[166]	43%[166]	45%[166]	41.28%[169]
USC-L B	Ultra-supercritical	Lignite	Fuel oil	675MW	43.70%[166]	35%[166]	42%[169]

Table 44 Power plant description - Fuel consumption during start-ups

Plant ID	Coldest start		Warm start		Hottest start	
	Main fuel MWhth/MW	Ignition fuel MWhth/MW	Main fuel MWhth/MW	Ignition fuel MWhth/MW	Main fuel MWhth/MW	Ignition fuel MWhth/MW
UC-H small	2.0[124]	1.4[124]	1.5[124]	0.9[124]	1.2[124]	0.6[124]
UC-H B	2.0[124]	1.4[124]	1.5[124]	0.9[124]	1.2[124]	0.6[124]
UC-H R	2.0[124]	1.4[124]	1.5[124]	0.9[124]	1.2[124]	0.6[124]
SC-H	3.5[124]	2.4[124]	3.1[124]	1.9[124]	2[124]	0.96[124]
USC-H W	4	2.8	3.6	1.9	2.2	1.1
USC-H T	4.1	2.9	3.7	2	2.3	1.2
USC-H S	4.2	3	3.8	2.1	2.4	1.3
USC-H	4	2.8	3.6	1.9	2.2	1.1
UC-Lo	2.0[170]	1.4[170]	1.5[170]	0.9[170]	1.2[170]	0.6[170]
UC-L	2.0[170]	1.4[170]	1.5[170]	0.9[170]	1.2[170]	0.6[170]
SC-L B	3.5[170]	2.4[170]	3.1[170]	1.9[170]	2[170]	0.96[170]
SC-L SP	3.5[170]	2.4[170]	3.1[170]	1.9[170]	2[170]	0.96[170]
USC-L N	4.1	2.9	3.7	2	2.3	1.2
USC-L B	4	2.8	3.6	1.9	2.2	1.1

Table 45 Power plant description - Start-up duration

Plant ID	Coldest start duration	Warm start	Hottest start
UC-H small	10h	4h	2h
UC-H B	480min[166]	4h	120min[166]
UC-H R	6h	4h	2h
SC-H	7h	5h	3h
USC-H W	300min[166]	4h	66min[166]
USC-H T	300min[166]	4h	140min[166]
USC-H S	550min[166]	4.5h	190min[166]
USC-H	8h	5h	3h
UC-Lo	7h	5h	3h
UC-L	6h	4h	2h
SC-L B	360min[166]	4h	140min[166]
SC-L SP	6h	4h	2h
USC-L N	380min[166]	3.5h	2h
USC-L B	300min[166]	3h	80min[166]

Table 46 Power plant description - Start-up ramp-rate

Plant ID	Coldest start-up ramp rate	Warm start-up ramp rate	Hottest start-up ramp rate
UC-H small	1.5%/min[164]	1.75%/min[164]	2%/min[164]
UC-H B	4.5%/min[166]	4.75%/min[166]	5%/min[166]
UC-H R	1.5%/min[167]	1.75%/min[167]	2%/min[167]
SC-H	3.5%/min[164]	3.75%/min[164]	4%/min[164]
USC-H W	4.5%/min[166]	4.75%/min[166]	5%/min[166]
USC-H T	1.5%/min[166]	1.75%/min[166]	2%/min[166]
USC-H S	2.5%/min[166]	2.75%/min[166]	3%/min[166]
USC-H	6.5%/min	6.75%/min	7%/min
UC-Lo	0.75%/min[167]	1%/min[167]	1.3%/min[[167]
UC-L	0.75%/min[167]	1%/min[167]	1.3%/min[167]
SC-L B	3.5%/min[166]	3.75%/min[166]	4%/min[166]
SC-L SP	3.5%/min[166]	3.75%/min[166]	4%/min[166]
USC-L N	2.5%/min[166]	2.75%/min[166]	3%/min[166]
USC-L B	3.5%/min[166]	3.75%/min[166]	4%/min[166]

Table 47 Power plant description - Ramp-rate

Plant ID	Ramp-up rate	Ramp-down rate	Frequency control ramp-up rate	Frequency control ramp-down rate
UC-H small	2%/min[164]	2%/min[164]	2%/min	2%/min
UC-H B	5%/min[166]	5%/min[166]	5%/min	5%/min
UC-H R	2%/min[[167] 7%/min [173, 173]	2%/min[167]	2%/min	2%/min
SC-H	4%/min[164]	4%/min[164]	2%/min	2%/min
USC-H W	5%/min[166]	5%/min[166]	5%/min	5%/min
USC-H T	2%/min[166]	2%/min[166]	2%/min	2%/min
USC-H S	3%/min[166]	3%/min[166]	3%/min	3%/min
USC-H	7%/min	7%/min	7%/min	7%/min
UC-Lo	1.3%/min[167]	1.3%/min[167]	1.3%/min	1.3%/min
UC-L	1.3%/min[167]	1.3%/min[167]	1.3%/min	1.3%/min
SC-L B	4%/min[166]	4%/min[[166]	4%/min	4%/min
SC-L SP	4%/min[166]	4%/min[166]	4%/min	4%/min
USC-L N	3%/min[166]	3%/min[166]	3%/min	3%/min
USC-L B	4%/min[166]	4%/min[166]	4%/min	4%/min

Table 48 Power plant description - Start-up maintenance costs

Plant ID	Nominal Load MW	Coldest start maintenance costs in €	Warm start maintenance costs in €	Hottest start maintenance costs in €
UC-H small	250MW	36750	39250	23500
UC-H B	660MW[166]	69300	42900	38940
UC-H R	506MW[167]	53130	32890	29854
SC-H	800MW	83200	51200	43200
USC-H W	725MW	75400	46400	39150
USC-H T	660MW	68640	42240	35640
USC-H S	500MW	52000	32000	27000
USC-H	400MW	41600	25600	21600
UC-Lo	400MW	42000	26000	23600
UC-L	400MW	42000	26000	23600
SC-L B	858[166]	60840	37440	31590
SC-L SP	800MW[173]	83200	51200	43200
USC-L N	1100MW[166]	114400	70400	59400
USC-L B	675MW	70200	43200	36450

References

[1] IEA - International Energy Agency, "Key World Energy Statistics 2017," Paris, France, 2017. [Online] Available: https://www.iea.org/publications/freepublications/publication/key-world-energy-statistics.html. Accessed on: Nov. 27 2017.

[2] O. Publishing, *CO2 Emissions from Fuel Combustion 2017 -*. Paris: OECD Publishing, 2017.

[3] PIK- Potsdam Institute for Climate Impact Research, *Paris Reality Check*. [Online] Available: https://www.pik-potsdam.de/primap-live/.

[4] O. Edenhofer, Ed., Climate change 2014: Mitigation of climate change Working Group III contribution to the Fifth Assessment Report of the Intergovernmental Panel on Climate Change. New York NY: Cambridge University Press, 2014.

[5] eurostat, *Energy from renewable sources*. [Online] Available: http://ec.europa.eu/eurostat/web/energy/data/shares.

[6] C. Breyer *et al.*, "Solar photovoltaics demand for the global energy transition in the power sector," *Prog Photovolt Res Appl*, vol. 114, no. 6223, p. 7, 2017.

[7] Artelys, Armines, and Energies Demain, "A 100% renewable electricity mix," 2016. [Online] Available: http://www.ademe.fr/node/122931. Accessed on: Nov. 10 2017.

[8] Rat von Sachverständigen für Umweltfragen, *Wege zur 100% erneuerbaren Stromversorgung: Sondergutachten, Januar 2011*. Berlin: Erich Schmidt Verlag, 2011.

[9] B. P. Heard, B. W. Brook, T.M.L. Wigley, and C.J.A. Bradshaw, "Burden of proof: A comprehensive review of the feasibility of 100% renewable-electricity systems," *Renewable and Sustainable Energy Reviews*, vol. 76, pp. 1122–1133, 2017.

[10] Climate Action Tracker, "The Coal Gap: planned coal-fired power plants inconsistent with 2 ° C and threaten achievement of INDCs," 2015.

[11] Climate analytics, "A stress test for coal in Europe under the Paris agreement," Climate analytics, 2017. [Online] Available: www.climateanalytics.org/publications. Accessed on: Jan. 16 2018.

[12] J. Rockström *et al.*, "A roadmap for rapid decarbonization," (eng), *Science (New York, N.Y.)*, vol. 355, no. 6331, pp. 1269–1271, 2017.

[13] A. Kruse, E. Kunle, and M. Faulstich, "Strukturwandel der konventionellen Stromversorgung als gesellschaftliche Aufgabe," in *Management-Reihe Corporate Social Responsibility, CSR und Energiewirtschaft*, A. Hildebrandt and W. Landhäußer, Eds., 1st ed., Berlin, Heidelberg: Springer Berlin Heidelberg, 2016, pp. 103–122.

[14] L. Sloss, "Levelling the intermittency of renewables with coal," IEA Clean Coal Centre CCC/268, 2016.

[15] M. A. Gonzalez-Salazar, T. Kirsten, and L. Prchlik, "Review of the operational flexibility and emissions of gas- and coal-fired power plants in a future with growing renewables," *Renewable and Sustainable Energy Reviews*, vol. 82, pp. 1497–1513, 2018.

[16] M. A. Bucher, S. Delikaraoglou, K. Heussen, P. Pinson, and G. Andersson, "On quantification of flexibility in power systems," in *PowerTech*, pp. 1–6.

[17] J. H. Zhao, Z. Y. Dong, P. Lindsay, and K. P. Wong, "Flexible Transmission Expansion Planning With Uncertainties in an Electricity Market," *IEEE Trans. Power Syst.*, vol. 24, no. 1, pp. 479–488, 2009.

[18] A. Capasso, M. C. Falvo, R. Lamedica, S. Lauria, and S. Scalcino, "A new methodology for power systems flexibility evaluation," in *2005 IEEE Russia Power Tech*, St. Petersburg, Russia, 2005, pp. 1–6.

[19] P. Bresesti, A. Capasso, M. C. Falvo, and S. Lauria, "Power system planning under uncertainty conditions: Criteria for transmission network flexibility evaluation," in *2003 IEEE Bologna PowerTech*, Bologna, Italy, 2003, p. 6.

[20] M. Lu, Z. Y. Dong, and T. K. Saha, "Transmission expansion planning flexibility," in *2005 International Power Engineering Conference*, Singapore, 2005, 893-898 Vol. 2.

[21] F. Bouffard and M. Ortega-Vazquez, "The value of operational flexibility in power systems with significant wind power generation," in *IEEE Power and Energy Society general meeting, 2011: 24 - 29 July 2011, Detroit, MI, USA*, Detroit, MI, USA, 2011, pp. 1–5.

[22] J. Ma, V. Silva, R. Belhomme, D. S. Kirschen, and L. F. Ochoa, "Evaluating and Planning Flexibility in Sustainable Power Systems," *IEEE Trans. Sustain. Energy*, vol. 4, no. 1, pp. 200–209, 2013.

[23] A. M. Ross, D. H. Rhodes, and D. E. Hastings, "Defining changeability: Reconciling flexibility, adaptability, scalability, modifiability, and robustness for maintaining system lifecycle value," *Syst. Engin.*, vol. 11, no. 3, pp. 246–262, 2008.

[24] International Energy Agency; Organisation for Economic Co-operation and Development, *Harnessing variable renewables: A guide to the balancing challenge / International Energy Agency*. Paris: IEA, 2011.

[25] H. Kondziella and T. Bruckner, "Flexibility requirements of renewable energy based electricity systems – a review of research results and methodologies," *Renewable and Sustainable Energy Reviews*, vol. 53, pp. 10–22, 2016.

[26] M. Miller *et al.*, "Status Report on Power System Transformation: A 21st Century Power Partnership Report,"

[27] H. Holttinen *et al.*, "The Flexibility Workout: Managing Variable Resources and Assessing the Need for Power System Modification," *IEEE Power and Energy Mag.*, vol. 11, no. 6, pp. 53–62, 2013.

[28] Q. Wang and B.-M. Hodge, "Enhancing Power System Operational Flexibility With Flexible Ramping Products: A Review," *IEEE Trans. Ind. Inf.*, vol. 13, no. 4, pp. 1652–1664, 2017.

[29] A. Ulbig and G. Andersson, "Analyzing Operational Flexibility of Power Systems," in International Journal of Electrical Power, The Special Issue for 18th Power Systems Computation Conference, Pages 155–164.

[30] B. F. Hobbs, J. C. Honious, and J. Bluestein, "Estimating the flexibility of utility resource plans: An application to natural gas cofiring for SO2 control," *IEEE Trans. Power Syst.*, vol. 9, no. 1, pp. 167–173, 1994.

[31] North American Electric Reliability Corporation, "Potential Reliability Impacts of Emerging Flexible Resources: NERC IVGTF Task 1-5," 2010.

[32] North American Electric Reliability Corporation, "Special report flexibility requirements and potential metrics for variable generation: Implications for system planning studies," NERC, Princeton, NJ, 2010.

[33] Y. V. Makarov, C. Loutan, J. Ma, and P. de Mello, "Operational Impacts of Wind Generation on California Power Systems," *IEEE Trans. Power Syst.*, vol. 24, no. 2, pp. 1039–1050, 2009.

[34] A. Ulbig and G. Andersson, "On operational flexibility in power systems," in *IEEE Power and Energy Society general meeting, 2012: 22 - 26 July 2012, San Diego, CA, USA*, San Diego, CA, 2012, pp. 1–8.

[35] E. Lannoye, D. Flynn, and M. O'Malley, "Evaluation of Power System Flexibility," pp. 922–931.

[36] T. Zheng, J. Zhao, F. Zhao, and E. Litvinov, "Operational flexibility and system dispatch," in *IEEE Power and Energy Society general meeting, 2012: 22 - 26 July 2012, San Diego, CA, USA*, San Diego, CA, 2012, pp. 1–3.

[37] N. Menemenlis, M. Huneault, and A. Robitaille, "Thoughts on power system flexibility quantification for the short-term horizon," in *IEEE Power and Energy Society general meeting, 2011: 24 - 29 July 2011, Detroit, MI, USA*, Detroit, MI, USA, 2011, pp. 1–8.

[38] H. Mangesius, S. Hirche, M. Huber, and T. Hamacher, Eds., A framework to quantify technical flexibility in power systems based on reliability certificates, 2013.

[39] K. Studarus and R. D. Christie, "A deterministic metric of stochastic operational flexibility," in *IEEE Power and Energy Society general meeting (PES), 2013: 21 - 25 July 2013, Vancouver, BC, Canada*, Vancouver, BC, 2013, pp. 1–4.

[40] M. K. Petersen, K. Edlund, L. H. Hansen, J. Bendtsen, and J. Stoustrup, "A taxonomy for modeling flexibility and a computationally efficient algorithm for dispatch in Smart Grids," in *American Control Conference (ACC), 2013: 17 - 19 June 2013, Washington, DC, USA*, Washington, DC, 2013, pp. 1150–1156.

[41] T. von Danwitz, "Regulation and liberalization of the European electricity market – a German view," in *Energy Law Journal,*, p. 423.

[42] M. Huber, D. Dimkova, and T. Hamacher, "Integration of wind and solar power in Europe: Assessment of flexibility requirements," *Energy*, vol. 69, pp. 236–246, 2014.

[43] J. Zhao, T. Zheng, and E. Litvinov, "A Unified Framework for Defining and Measuring Flexibility in Power System," *IEEE Trans. Power Syst.*, vol. 31, no. 1, pp. 339–347, 2016.

[44] P. L. Joskow, "Lessons learned from electricity market liberalization," in *Energy Journal*, pp. 9–42.

[45] A. Zervos and C. Lins, *Renewables 2016 Global Status Report*. Ottawa, ON, CA: REN21, 2016.

[46] R. Baldick, U. Helman, B. F. Hobbs, and R. P. O'Neill, "Design of Efficient Generation Markets," *Proc. IEEE*, vol. 93, no. 11, pp. 1998–2012, 2005.

[47] M. Glowacki, *European Union Electricity Market Glossary.* [Online] Available: https://www.emissions-euets.com/internal-electricity-market-glossary. Accessed on: Oct. 26 2017.

[48] P. N. Biskas *et al.*, "High-level design for the compliance of the Greek wholesale electricity market with the Target Model provisions in Europe," *Electric Power Systems Research*, vol. 152, pp. 323–341, 2017.

[49] D. R. Biggar and M. Hesamzadeh, *The economics of electricity markets.* Chichester: Wiley, 2014.

[50] T. Jamasb and M. Pollitt, "Electricity Market Reform in the European Union: Review of Progress toward Liberalization & Integration," in *The Energy Journal Special Issue 2005*, pp. 11–41.

[51] Agora Energiewende, "Flexibility in thermal power plants: With a focus on existing coal-fired power plants," 2017.

[52] C. Kiyak and A. de Vries, "Electricity Market Mechanism regarding the Operational Flexibility of Power Plants," *ME*, vol. 08, no. 04, pp. 567–589, 2017.

[53] PLEF, "Generation Adequacy Assessment 2018," 2018. [Online] Available: http://www.bmwi.de/Redaktion/DE/Downloads/P-R/plef-sg2-generation-adequacy-assessment-2018.pdf?__blob=publicationFile&v=4. Accessed on: Feb. 07 2018.

[54] M. Fornacciari, C. July, and M. Motylewski, Capacity Markets — A short-term fix for security of supply or a key energy market support mechanism in the transition to a low carbon economy? [Online] Available: https://www.dentons.com/en/insights/articles/2014/october/28/capacity-mar-kets?utm_source=Mondaq&utm_medium=syndication&utm_campaign=View-Original. Accessed on: Feb. 07 2018.

[55] E. Lannoye, D. Flynn, and M. O'Malley, "Power system flexibility assessment — State of the art," in *IEEE Power and Energy Society general meeting, 2012: 22 - 26 July 2012, San Diego, CA, USA*, San Diego, CA, 2012, pp. 1–6.

[56] H. Holttinen, Design and operation of power systems with large amounts of wind power: Final report, IEA WIND Task 25, phase one 2006-2008. Espoo: VTT, 2009.

[57] E. Lannoye, D. Flynn, and M. O'Malley, "The role of power system flexibility in generation planning," in *IEEE Power and Energy Society general meeting, 2011: 24 - 29 July 2011, Detroit, MI, USA*, Detroit, MI, USA, 2011, pp. 1–6.

[58] Tuohy and Lannoye, "Metrics for Quantifying Flexibility in Power System Planning," Electric Power Research Institute (EPRI),, Palo Alto, California.

[59] A. A. Thatte and Le Xie, "A Metric and Market Construct of Inter-Temporal Flexibility in Time-Coupled Economic Dispatch," *IEEE Trans. Power Syst.*, vol. 31, no. 5, pp. 3437–3446, 2016.

[60] A.M.L.L. da Silva, W. S. Sales, L. A. da Fonseca Manso, and R. Billinton, "Long-Term Probabilistic Evaluation of Operating Reserve Requirements With Renewable Sources," *IEEE Trans. Power Syst.*, vol. 25, no. 1, pp. 106–116, 2010.

[61] Y. Dvorkin, D. S. Kirschen, and M. A. Ortega-Vazquez, "Assessing flexibility requirements in power systems," *IET Generation, Transmission & Distribution*, vol. 8, no. 11, pp. 1820–1830, 2014.

[62] J. Bertsch, C. Growitsch, S. Lorenczik, and S. Nagl, "Flexibility in Europe's power sector — An additional requirement or an automatic complement?," *Energy Economics*, vol. 53, pp. 118–131, 2016.

[63] Byfield and Vetter, Eds., "Flexibility concepts for the German power supply in 2050: Ensuring stability in the age of renewable energies," acatech, 2016.

[64] European Commission, "Mainstreaming RES Flexibility portfolios Design of flexibility portfolios at Member State level to facilitate a cost-efficient integration of high shares of renewables," Jul. 2017. [Online] Available: https://ec.europa.eu/energy/sites/ener/files/mainstreaming_res_-_artelys_-_final_report_-_version_33.pdf. Accessed on: Oct. 05 2017.

[65] Yasuda *et al.*, "Flexibility Chart: Evaluation on diversity of flexibility in various areas,"

[66] A. Zangeneh, S. Jadid, and A. Rahimi-Kian, "Uncertainty based distributed generation expansion planning in electricity markets," *Electr Eng*, vol. 91, no. 7, pp. 369–382, 2010.

[67] D. Mejia-Giraldo and J. D. McCalley, "Maximizing Future Flexibility in Electric Generation Portfolios," *IEEE Trans. Power Syst.*, vol. 29, no. 1, pp. 279–288, 2014.

[68] M. A. Bucher, S. Chatzivasileiadis, and G. Andersson, "Managing Flexibility in Multi-Area Power Systems," *IEEE Trans. Power Syst.*, vol. 31, no. 2, pp. 1218–1226, 2016.

[69] E. Lannoye, D. Flynn, and M. O'Malley, "Transmission, Variable Generation, and Power System Flexibility," *IEEE Trans. Power Syst.*, vol. 30, no. 1, pp. 57–66, 2015.

[70] S. Borenstein, *The Private and Public Economics of Renewable Electricity Generation*. Cambridge, MA: National Bureau of Economic Research, 2011.

[71] H. K. H. Ommedal, "Cost of flexibility in the future European power system," Master Thesis, Department of Electric Power Engineering, NTNU Norwegian University of Science and Technology, 2015.

[72] IEA - International Energy Agency, *The power of transformation: Wind, sun and the economics of flexible power systems*. Paris, France: IEA - International Energy Agency, 2014.

[73] EIA, "Levelized Cost and Levelized Avoided Cost of New Generation Resources in the Annual Energy Outlook 2017," AEO2017, 2017.

[74] L. Hirth, F. Ueckerdt, and O. Edenhofer, "Why Wind Is Not Coal: On the Economics of Electricity Generation," *EJ*, vol. 37, no. 3, 2016.

[75] P. L. Joskow, "Comparing the Costs of Intermittent and Dispatchable Electricity Generating Technologies," *The American Economic Review*, vol. 101, no. 3, pp. 238–241, http://www.jstor.org/stable/29783746, 2011.

[76] Q. Wang, H. Wu, A. R. Florita, C. Brancucci Martinez-Anido, and B.-M. Hodge, "The value of improved wind power forecasting: Grid flexibility quantification, ramp capability analysis, and impacts of electricity market operation timescales," *Applied Energy*, vol. 184, pp. 696–713, 2016.

[77] D. S. Kirschen, J. Ma, V. Silva, and R. Belhomme, "Optimizing the flexibility of a portfolio of generating plants to deal with wind generation," in *IEEE Power*

and *Energy Society general meeting, 2011: 24 - 29 July 2011, Detroit, MI, USA*, Detroit, MI, USA, 2011, pp. 1–7.

[78] S. Lu *et al.*, "Unit commitment considering generation flexibility and environmental constraints," in *IEEE PES General Meeting*, Minneapolis, MN, 2010, pp. 1–11.

[79] FfE, "Merit Order der Energiespeicherung im Jahr 2030," Forschungsstelle für Energiewirtschaft e.V., 2016.

[80] D. T. Gardner, "Flexibility in electric power planning: Coping with demand uncertainty," *Energy*, vol. 21, no. 12, pp. 1207–1218, 1996.

[81] A. S. Brouwer, M. van den Broek, A. Seebregts, and A. Faaij, "Operational flexibility and economics of power plants in future low-carbon power systems," *Applied Energy*, vol. 156, pp. 107–128, 2015.

[82] E. Lannoye *et al.*, "Integration of variable generation: Capacity value and evaluation of flexibility," in *IEEE PES General Meeting*, Minneapolis, MN, 2010, pp. 1–6.

[83] J. Hentschel, U. Babić, and H. Spliethoff, "A parametric approach for the valuation of power plant flexibility options," *Energy Reports*, vol. 2, pp. 40–47, 2016.

[84] J. Kopiske, S. Spieker, and G. Tsatsaronis, "Value of power plant flexibility in power systems with high shares of variable renewables: A scenario outlook for Germany 2035," *Energy*, vol. 137, pp. 823–833, 2017.

[85] R. Loisel, A. Mercier, C. Gatzen, N. Elms, and H. Petric, "Valuation framework for large scale electricity storage in a case with wind curtailment," *Energy Policy*, vol. 38, no. 11, pp. 7323–7337, 2010.

[86] C. O'Dwyer and D. Flynn, "Using Energy Storage to Manage High Net Load Variability at Sub-Hourly Time-Scales," *IEEE Trans. Power Syst.*, vol. 30, no. 4, pp. 2139–2148, 2015.

[87] E. Ela and M. O'Malley, "Studying the Variability and Uncertainty Impacts of Variable Generation at Multiple Timescales," *IEEE Trans. Power Syst.*, vol. 27, no. 3, pp. 1324–1333, 2012.

[88] J. P. Deane, G. Drayton, and B. P. Ó Gallachóir, "The impact of sub-hourly modelling in power systems with significant levels of renewable generation," *Applied Energy*, vol. 113, pp. 152–158, 2014.

[89] M. I. Alizadeh, M. Parsa Moghaddam, N. Amjady, P. Siano, and M. K. Sheikh-El-Eslami, "Flexibility in future power systems with high renewable penetration: A review," *Renewable and Sustainable Energy Reviews*, vol. 57, pp. 1186–1193, 2016.

[90] B. F. Hobbs, M. H. Rothkopf, R. P. O'Neill, and H.-p. Chao, Eds., *The Next Generation of Electric Power Unit Commitment Models*. Boston, MA: Kluwer Academic Publishers, 2002.

[91] S. Sen and D. P. Kothari, "Optimal thermal generating unit commitment: A review," pp. 443–451.

[92] M. Tahanan, W. van Ackooij, A. Frangioni, and F. Lacalandra, "Large-scale Unit Commitment under uncertainty," *4OR-Q J Oper Res*, vol. 13, no. 2, pp. 115–171, 2015.

[93] N. P. Padhy, "Unit Commitment—A Bibliographical Survey," *IEEE Trans. Power Syst.*, vol. 19, no. 2, pp. 1196–1205, 2004.

[94] N. M. Pindoriya, S. N. Singh, and J. Ø stergaard, "Day-Ahead Self-Scheduling of Thermal Generator in Competitive Electricity Market Using Hybrid PSO," pp. 1–6.

[95] G. Morales-Espana, J. M. Latorre, and A. Ramos, "Tight and Compact MILP Formulation of Start-Up and Shut-Down Ramping in Unit Commitment," pp. 1288–1296.

[96] M. Ventosa, Á. Baíllo, A. Ramos, and M. Rivier, "Electricity market modeling trends," *Energy Policy*, vol. 33, no. 7, pp. 897–913, 2005.

[97] Gross and Finlay, "Optimal bidding strategies in competitive electricity markets," in *Proceedings of the 12th PSCC,1996*.

[98] G. B. Sheble and G. N. Fahd, "Unit commitment literature synopsis," *IEEE Trans. Power Syst.*, vol. 9, no. 1, pp. 128–135, 1994.

[99] J. Campion, C. Dent, M. Fox, D. Long, and D. Magazzeni, "Challenge: Modelling Unit Commitment as a Planning Problem," in *Proceedings of the Twenty-Third International Conference on Automated Planning and Scheduling*.

[100] J. Zhu, *Optimization of power system operation*. Hoboken: IEEE Press/Wiley, 2015.

[101] L. H. Fink, "Discrete events in Power Systems," *Discrete Event Dynamic Systems*, vol. 9, no. 4, pp. 319–330, 1999.

[102] Robert E. Bixby, "A Brief History of Linear and Mixed-Integer Programming Computation," in *DOCUMENTA MATHEMATICA*, pp. 107–121.

[103] G. Morales-Espana, J. M. Latorre, and A. Ramos, "Tight and compact MILP formulation of start-up and shut-down ramping in unit commitment," in *IEEE Power and Energy Society general meeting (PES), 2013: 21 - 25 July 2013, Vancouver, BC, Canada*, Vancouver, BC, 2013, p. 1.

[104] M. Carrion and J. M. Arroyo, "A Computationally Efficient Mixed-Integer Linear Formulation for the Thermal Unit Commitment Problem," *IEEE Trans. Power Syst.*, vol. 21, no. 3, pp. 1371–1378, 2006.

[105] A. Frangioni, C. Gentile, and F. Lacalandra, "Tighter Approximated MILP Formulations for Unit Commitment Problems," *IEEE Trans. Power Syst.*, vol. 24, no. 1, pp. 105–113, 2009.

[106] Lima and Grossmann, "Computational advances in solving Mixed Integer Linear Programming problems,"

[107] R. Shoults, S. Chang, S. Helmick, and W. Grady, "A Practical Approach to Unit Commitment, Economic Dispatch and Savings Allocation for Multiple-Area Pool Operation with Import/Export Constraints," *IEEE Trans. on Power Apparatus and Syst.*, vol. PAS-99, no. 2, pp. 625–635, 1980.

[108] P. Lowery, "Generating Unit Commitment by Dynamic Programming," *IEEE Trans. on Power Apparatus and Syst.*, vol. PAS-85, no. 5, pp. 422–426, 1966.

[109] A. I. Cohen and M. Yoshimura, "A Branch-and-Bound Algorithm for Unit Commitment," *IEEE Power Eng. Rev.*, vol. PER-3, no. 2, pp. 34–35, 1983.

[110] M. L. Fisher, "Optimal Solution of Scheduling Problems Using Lagrange Multipliers: Part I," *Operations Research*, vol. 21, no. 5, pp. 1114–1127, 1973.

[111] J. Waight, F. Albuyeh, and A. Bose, "Scheduling of Generation and Reserve Margin Using Dynamic and Linear Programming," *IEEE Trans. on Power Apparatus and Syst.*, vol. PAS-100, no. 5, pp. 2226–2230, 1981.

[112] C. A. Floudas and X. Lin, "Mixed Integer Linear Programming in Process Scheduling: Modeling, Algorithms, and Applications," *Annals of Operations Research*, vol. 139, no. 1, pp. 131–162, 2005.

[113] Fox, Long, and Magazzeni, "Plan-based Policies for Efficient Multiple Battery Load Management," *Journal Of Artificial Intelligence Research*, vol. 44, pp. 335–382, 2012.

[114] Piacentini, Magazzeni, Long, Fox, and Dent, "Solving realistic unit commitment problems using temporal planning: challenges and solutions," in *Proceedings of the Twenty-Sixth International Conference on Automated Planning and Scheduling*, 2016, pp. 421–430.

[115] C. G. Cassandras and S. Lafortune, *Introduction to discrete event systems*, 2nd ed. New York, N.Y: Springer Science+Business Media, 2008.

[116] J. E. Hopcroft, R. Motwani, and J. D. Ullman, *Introduction to automata theory, languages and computation*, 3rd ed. Boston, MA: Pearson/Addison Wesley, 2007.

[117] Y. Abdeddaïm, E. Asarin, and O. Maler, "Scheduling with timed automata," *Theoretical Computer Science*, vol. 354, no. 2, pp. 272–300, 2006.

[118] S. Di Cairano, A. Bemporad, and J. Júlvez, "Event-driven optimization-based control of hybrid systems with integral continuous-time dynamics," *Automatica*, vol. 45, no. 5, pp. 1243–1251, 2009.

[119] P. Niemczyk, "Model-based Fuel Flow Control for Fossil-fired Power Plants," PhD, Aalborg University, 2010.

[120] X. D. Koutsoukos, K. X. He, M. D. Lemmon, and P. J. Antsaklis, "Timed Petri Nets in Hybrid Systems: stability and supervisory Control," *Discrete Event Dynamic Systems*, vol. 8, no. 2, pp. 137–173, 1998.

[121] J. P. Deane, A. Chiodi, M. Gargiulo, and B. P. Ó Gallachóir, "Soft-linking of a power systems model to an energy systems model," *Energy*, vol. 42, no. 1, pp. 303–312, 2012.

[122] K. Helbig, M. Banaszkiewicz, and W. Mohr, "Advanced lifetime assessment and stress control of steam turbines," in *PowerGen Europe, Milano 2016*.

[123] J. Vogt, T. Schaaf, H. Sun, and K. Helbig, "Impact of Low Load Operation on Plant Lifetime," in Proceedings of the ASME Turbo Expo: Turbine Technical Conference and Exposition 2014 : presented at the ASME 2014 Turbo Expo: Turbine Technical Conference and Exposition, June 16-20, 2014, Düsseldorf, Germany, Düsseldorf, Germany, 2014, V01BT27A026.

[124] N. Kumar, P. Besuner, Lefton S, D. Agan, and D. Hilleman, "Power Plant Cycling Costs," NREL, Intertek APTECH Sunnyvale, California,, Tech. Rep. NREL SR-5500-55433, Apr. 2012. [Online] Available: http://www.osti.gov/bridge.

[125] Maher *et al.*, *SCIP*: Zuse Institute Berlin, 2017.

[126] Maher *et al.*, "The SCIP Optimization Suite 4.0," 2017. [Online] Available: https://opus4.kobv.de/opus4-zib/frontdoor/index/index/docId/6217. Accessed on: Nov. 09 2017.

[127] G. L. Nemhauser and L. A. Wolsey, *Integer and combinatorial optimization*. New York, Chichester: Wiley, 1999.

[128] M. Sniedovich, Dynamic programming: Foundations and principles / Moshe Sniedovich, 2nd ed. New York: CRC, 2011.

[129] B. Saravanan, S. Das, S. Sikri, and D. P. Kothari, "A solution to the unit commitment problem—a review," *Front. Energy*, vol. 7, no. 2, pp. 223–236, 2013.

[130] A. Bhardwaj, V. K. Kamboj, V. K. Shukla, B. Singh, and P. Khurana, "Unit commitment in electrical power system-a literature review," in *2012 IEEE International Power Engineering and Optimization Conference*, Melaka, Malaysia, 2012, pp. 275–280.

[131] D. K. Kaynar, *The theory of timed I/O Automata*, 2nd ed. San Rafael, Calif.: Morgan & Claypool, 2011.

[132] J. Garcia-Gonzalez, A. M. San Roque, F. A. Campos, and J. Villar, "Connecting the Intraday Energy and Reserve Markets by an Optimal Redispatch," *IEEE Trans. Power Syst.*, vol. 22, no. 4, pp. 2220–2231, 2007.

[133] L. Hirth and I. Ziegenhagen, "Balancing power and variable renewables: Three links," pp. 1035–1051.

[134] G. Morales-Espana, R. Baldick, J. Garcia-Gonzalez, and A. Ramos, "Power-Capacity and Ramp-Capability Reserves for Wind Integration in Power-Based UC," *IEEE Trans. Sustain. Energy*, vol. 7, no. 2, pp. 614–624, 2016.

[135] H. Wu, M. Shahidehpour, and M. E. Khodayar, "Hourly Demand Response in Day-Ahead Scheduling Considering Generating Unit Ramping Cost," *IEEE Trans. Power Syst.*, vol. 28, no. 3, pp. 2446–2454, 2013.

[136] H. Wu, M. Shahidehpour, A. Alabdulwahab, and A. Abusorrah, "Thermal Generation Flexibility With Ramping Costs and Hourly Demand Response in Stochastic Security-Constrained Scheduling of Variable Energy Sources," *IEEE Trans. Power Syst.*, vol. 30, no. 6, pp. 2955–2964, 2015.

[137] E. Centeno et al., "A Goal Programming Model for Rescheduling of Generation Power in Deregulated Markets," *Annals of Operations Research*, vol. 120, no. 1/4, pp. 45–57, 2003.

[138] J. M. Arroyo and A. J. Conejo, "Optimal response of a power generator to energy, AGC, and reserve pool-based markets," *IEEE Trans. Power Syst.*, vol. 17, no. 2, pp. 404–410, 2002.

[139] C. K. Simoglou, P. N. Biskas, and A. G. Bakirtzis, "Optimal Self-Scheduling of a Thermal Producer in Short-Term Electricity Markets by MILP," *IEEE Trans. Power Syst.*, vol. 25, no. 4, pp. 1965–1977, 2010.

[140] G. Morales-Espana, A. Ramos, and J. Garcia-Gonzalez, "An MIP Formulation for Joint Market-Clearing of Energy and Reserves Based on Ramp Scheduling," *IEEE Trans. Power Syst.*, vol. 29, no. 1, pp. 476–488, 2014.

[141] Ocker, Ehrhart, and Ott, "An economic analysis of the German secondary balancing power market," 2015. [Online] Available: http://games.econ.kit.edu/img/Economic_Analysis_SR.pdf. Accessed on: Nov. 13 2017.

[142] F. S. Wen and A. K. David, "Optimally co-ordinated bidding strategies in energy and ancillary service markets," *IEE Proc., Gener. Transm. Distrib.*, vol. 149, no. 3, p. 331, 2002.

[143] T. Wu, M. Rothleder, Z. Alaywan, and A. D. Papalexopoulos, "Pricing Energy and Ancillary Services in Integrated Market Systems by an Optimal Power Flow," *IEEE Trans. Power Syst.*, vol. 19, no. 1, pp. 339–347, 2004.

[144] F. D. Galiana, F. Bouffard, J. M. Arroyo, and J. F. Restrepo, "Scheduling and Pricing of Coupled Energy and Primary, Secondary, and Tertiary Reserves," *Proc. IEEE*, vol. 93, no. 11, pp. 1970–1983, 2005.

[145] R. Englert and F. Wortmann, "Proposed Roadmap for Implementing Automatic Generation Control (AGC): Report by the Indo-German Energy Programme Green Energy Corridors," May. 2016. [Online] Available: https://energyforum.in/publication-show/items/IGEP-Green_Energy_Corridor.html. Accessed on: Oct. 04 2017.

[146] Richter, "On the interaction between product markets and markets for production capacity: The case of the electricity industry," Institute of Energy Economics at the University of Cologne (EWI), 2012.

[147] F. Müsgens, A. Ockenfels, and M. Peek, "Economics and design of balancing power markets in Germany," *International Journal of Electrical Power & Energy Systems*, vol. 55, pp. 392–401, 2014.

[148] S. Just and C. Weber, "Pricing of reserves: Valuing system reserve capacity against spot prices in electricity markets," *Energy Economics*, vol. 30, no. 6, pp. 3198–3221, 2008.

[149] EPEX SPOT, New ID 3 -Price Index on German Intraday continuous market, 2015.

[150] Great Britain, Ofgem's retail market review: Sixth report of session 2010-12. London: Stationery Office, 2011.

[151] Merino and Ebrill, "Market Monitoring Report 2016 - Electricity: Electricity Wholesale Markets Volume," ACER; CEER, Jun. 2017.

[152] European Commission, "Monitoring progress towards the Energy Union objectives – key indicators: Second Report on the State of the Energy Union," European Commission, Brussels, 2017. [Online] Available: https://ec.europa.eu/commission/sites/beta-political/files/swd-energy-union-key-indicators_en.pdf. Accessed on: Jan. 14 2018.

[153] M. Waldron and Y. Nobuoka, Commentary: Changing utility business models and electricity investment in Europe, 2017.

[154] V. Grimm, A. Ockenfels, and G. Zoettl, "Strommarktdesign: Zur Ausgestaltung der Auktionsregeln an der EEX," *Zeitschrift für Energiewirtschaft*, vol. 32, no. 3, pp. 147–161, https://doi.org/10.1007/s12398-008-0020-7, 2008.

[155] M. Burger, B. Graeber, and G. Schindlmayr, Managing energy risk: An integrated view on power and other energy markets / Markus Burger, Bernhard Graeber, Gero Schindlmayr. Chichester: John Wiley, 2007.

[156] F. Roques, "Technology Choices for New Entrants in Liberalised Markets: The Value of Operating Flexibility and Contractual Arrangements," Cambridge Working Papers in Economics, Faculty of Economics, University of Cambridge, 2007. [Online] Available: https://EconPapers.repec.org/RePEc:cam:camdae:0759.

[157] Q. P. Zheng, J. Wang, and A. L. Liu, "Stochastic Optimization for Unit Commitment—A Review," *IEEE Trans. Power Syst.*, vol. 30, no. 4, pp. 1913–1924, 2015.

[158] J. Dupa?ov, N. Grwe-Kuska, and W. Rmisch, "Scenario reduction in stochastic programming," *Mathematical Programming*, vol. 95, no. 3, pp. 493–511, 2003.

[159] N. Growe-Kuska, H. Heitsch, and W. Romisch, "Scenario reduction and scenario tree construction for power management problems," in *2003 IEEE Bologna PowerTech*, Bologna, Italy, 2003, pp. 152–158.

[160] entsoe, *entsoe Transparency Platform: Actual generation per generation unit [16.1.A]*. [Online] Available: https://transparency.entsoe.eu/. Accessed on: Feb. 05 2018.

[161] S. Venkataraman *et al.*, "Cost-Benefit Analysis of Flexibility Retrofits for Coal and Gas-Fueled Power Plants," NREL, Aug. 2012 - Dec. 2013. [Online] Available: https://www.nrel.gov/docs/fy14osti/60862.pdf. Accessed on: Feb. 01 2018.

[162] H. Spliethoff, *Power generation from solid fuels*. Heidelberg: Springer, 2010.

[163] C. Henderson, "Upgrading and efficiency improvement in coal-fired power plants, CCC/221," 2013.

[164] C. Henderson, "Increasing the flexibility of coal -fired power plants," 2014.

[165] K. van den Bergh and E. Delarue, "Cycling of conventional power plants: Technical limits and actual costs," *Energy Conversion and Management*, vol. 97, pp. 70–77, 2015.

[166] Clerens, Farley, Jazbec, Kraus, and Tigges, "Thermal power in 2030: Added value for EU energy policy," European Power Plant Suppliers Association,

[167] C. Ziems, S. Meinke, and J. Nocke, "FE333 Langfassung - Kraftwerksbetrieb bei Einspeisung von Windparks und Photovoltaikanlagen: Untersuchung bezüglich der Auswirkungen von fluktuierender Windenergie- und Photovoltaikeinspeisung auf den Kraftwerkseinsatz sowie das regel- und thermodynamische Betriebsverhalten konventioneller Kraftwerke in Deutschland - ein vertiefter Blick auf die Veränderungen bis zum Jahr 2023 und die Ableitung künftiger Anforderungen," 2012. [Online] Available: https://www.vgb.org/vgbmultimedia/333_Abschlussbericht-p-5968.pdf. Accessed on: Nov. 22 2017.

[168] M. Kopac and A. Hilalci, "Effect of ambient temperature on the efficiency of the regenerative and reheat Çatalağzı power plant in Turkey," *Applied Thermal Engineering*, vol. 27, no. 8-9, pp. 1377–1385, 2007.

[169] IEA - International Energy Agency and Coal Industry Advisory Board, Power Generation from Coal: Measuring and Reporting Efficiency Performance and CO2 Emissions: OECD/IEA, 2010.

[170] W.-P. Schill, M. Pahle, and C. Gambardella, "On Start-up Costs of Thermal Power Plants in Markets with Increasing Shares of Fluctuating Renewables," Discussion Papers of DIW Berlin, DIW Berlin, German Institute for Economic Research 1540, 2016. [Online] Available: https://ideas.repec.org/p/diw/diwwpp/dp1540.html.

[171] G. Milojcic and Y. Dyllong, "Vergleich der Flexibilität und der CO2-Emissionen von Kohlen- und Gaskraftwerken," ENERGIEWIRTSCHAFTLICHE TAGESFRAGEN, Jul. 2016. [Online] Available: http://www.et-energie-

online.de/Portals/0/PDF/zukunftsfragen_2016_07_milojcic.pdf. Accessed on: Nov. 22 2017.

[172] P. Kokopeli, J. Schreifels, and R. Forte, "Assessment of startup period at coal-fired electric generating units - Revised," 2014. [Online] Available: https://www3.epa.gov/airtoxics/utility/matsssfinalruletsd110414.pdf. Accessed on: Nov. 22 2017.

[173] G. Heimann, "Adapting lignite power plants to new market conditions: VGB-igef Workshop "It's all about Flexibility"," Vattenfall EUrope Generation AG, 2016. [Online] Available: http://www.eqmagpro.com/adapting-lignite-power-plants-to-new-market-conditions/. Accessed on: Nov. 22 2017.

[174] J. Lindsay and K. Dragoon, "RNP Summary Report on Coal Plant Dynamic Performance Capability," RNP, 2010. [Online] Available: http://www.rnp.org/node/rnp-report-coal-plant-performance-capability. Accessed on: Jan. 22 2018.

[175] P. Keatley, A. Shibli, and N. J. Hewitt, "Estimating power plant start costs in cyclic operation," *Applied Energy*, vol. 111, pp. 550–557, 2013.

[176] National Renewable Energy Laboratory; Black & Veatch Building a world of difference, "Cost and performance data for power generation technologies," 2012. [Online] Available: https://www.bv.com/docs/reports-studies/nrel-cost-report.pdf. Accessed on: Nov. 22 2017.

[177] J. Trüby, "Thermal Power Plant Economics and Variable Renewable Energies: A Model-based Case Study for Germany," IEA - International Energy Agency; OECD, 2014. [Online] Available: https://www.iea.org/publications/insights/insightpublications/thermal-power-plant-economics-and-variable-renewable-energies.html. Accessed on: Nov. 22 2017.

[178] Danish Energy Agency, "Regulation and planning of district heating in Denmark," Danish Energy Agency, 2017. [Online] Available: https://ens.dk/sites/ens.dk/files/Globalcooperation/regulation_and_planning_of_district_heating_in_denmark.pdf. Accessed on: Jan. 09 2018.

[179] P. Bullinger, "Enhanced water/steam cycle for advanced combined cycle technology," in *Power Gen Asia 2012*.

[180] A. Carrino and R. B. Jones, "Modeling new coal projects: supercritical or subcritical?," in *Power Engineering 2006*.

[181] FDBR, "Anpassung thermischer Kraftwerke an künftigen Strommarkt: Technische Lösungen für die gesamte Technologiekette im Kraftwerk," 2013. [Online] Available: http://fdbr.kmw.firma.cc/index.php?id=358. Accessed on: Nov. 20 2017.

[182] Mitsubishi Hitachi Power Systems Europe, "Modernization of Combustion Systems," Mitsubishi Hitachi Power Systems Europe. [Online] Available: http://www.eu.mhps.com/media/files/broschueren/technologie/GB_Prsp_Modernisierung_Grossdampferzeuger_3-2014_SCREEN.pdf. Accessed on: Nov. 20 2017.

[183] IEA - International Energy Agency, "Status of Power System Transformation 2017," 2017.

[184] V. S, G. G P, and K. Singhania, "Role of Renewable Energy in Indian Power Sector," *Energy Procedia*, vol. 138, pp. 1073–1078, 2017.

[185] ENTSO-E WGAS, "Survey on ancillary services procurement, balancing market design 2016," 2017. [Online] Available: https://www.entsoe.eu/publications/market-reports/ancillary-services-survey/Pages/default.aspx. Accessed on: Dec. 18 2017.

[186] RTE, *RTE Customer's area - Ancillary services*. [Online] Available: http://clients.rte-france.com/lang/an/clients_producteurs/services_clients/services_systeme.jsp. Accessed on: Nov. 29 2017.

[187] RTE, *RTE Customer's area - Rapid and Complementary Reserve*. [Online] Available: https://clients.rte-france.com/lang/an/clients_producteurs/services_clients/reserves_rc.jsp. Accessed on: Nov. 29 2017.

[188] Energinet, Ed., "Ancillary services to be delivered in Denmark- tender conditions: Valid from 1 September 2017," 2017. [Online] Available: http://financedocbox.com/Insurance/65982431-Ancillary-services-to-be-delivered-in-denmark-tender-conditions.html. Accessed on: Dec. 19 2017.

[189] European Commission, "METIS Technical Note T4: Overview of European Electricity Markets," 2016.

[190] ree-Red Electrica de Espana, "Ancillary Services. Preliminary Report 2016," Madrid, Feb. 2017. [Online] Available: http://www.ree.es/sites/default/files/downloadable/ancillary-services-preliminary-2016.pdf. Accessed on: Dec. 19 2017.

[191] destatis, "Data on energy price trends: Long-time series from January 2000 to July 2017," 2017. [Online] Available: https://www.destatis.de/DE/Publikationen/Thematisch/Preise/Energiepreise/EnergyPriceTrendsPDF_5619002.html. Accessed on: Nov. 22 2017.

[192] StatBank Denmark.dk. [Online] Available: http://www.dst.dk/en/Statistik/emner/priser-og-forbrug/forbrugerpriser. Accessed on: Jan. 10 2018.

[193] S. Cornot-Gandolphe, *Indian steam coal imports: The great equation*. Oxford: Oxford Institute for Energy Studies, 2016.

[194] Government of India, "Indian petroleum & natural gas statistics 2015-16," New Delhi, 2016. [Online] Available: http://petroleum.nic.in/sites/default/files/pngstat_1.pdf. Accessed on: Dec. 30 2017.

[195] World Bank and Ecofys, "Carbon Pricing Watch 2017," World Bank, Washington, DC, 2017. [Online] Available: https://openknowledge.worldbank.org/handle/10986/26565. Accessed on: Dec. 11 2017.

[196] World Bank Group, "State and Trends of Carbon Pricing 2017," Washington DC, 2017.

[197] Partnership for Market Readiness (PMR), "Carbon tax guide: A Handbook for Policy Makers," 2017.

[198] OECD, "Effective Carbon Rates - Pricing CO2 through Taxes and Emissions Trading Systems - en - OECD," [Online] Available: http://www.oecd.org/tax/tax-policy/effective-carbon-rates-9789264260115-en.htm#country_profiles. Accessed on: Dec. 11 2017.

[199] IEA - International Energy Agency, "Energy prices and taxes: Country notes," IEA - International Energy Agency, 2017.

[200] J. Devlin, K. Li, P. Higgins, and A. Foley, "Gas generation and wind power: A review of unlikely allies in the United Kingdom and Ireland," *Renewable and Sustainable Energy Reviews*, vol. 70, pp. 757–768, 2017.

[201] D. Newbery, M. Pollitt, R. Ritz, and W. Strielkowski, "Market design for a high-renewables European electricity system," Cambridge Working Papers in Economics, Faculty of Economics, University of Cambridge 1726, 2017. [Online] Available: https://ideas.repec.org/p/cam/camdae/1726.html.

[202] J. Winkler, "Electricity market design for 100% renewable electricity in Germany," in *12. Symposium Energieinnovation*.

[203] K. Salovaara, S. Honkapuro, M. Makkonen, and O. Gore, "100 % renewable energy system - challenges and opportunities for electricity market design," in *2016 13th International Conference on the European Energy Market (EEM): 6-9 June 2016, Porto, Portugal*, Porto, Portugal, 2016, pp. 1–5.